Issues in Economic Development

The Authors

Prof. Victor Adeyeye Abimbola graduated from University of Ife in B.Sc (honours) in agriculture in 1978 and M.Sc. in Agricultural Economics in 1980 from University of Ibadan, Nigeria. He earned Ph.D. degree in Agricultural Economics from University of Ibadan, Nigeria in year 1986. Professor V.A. Adeyeye has published more than 150 research papers in field of Agricultural Economics and Economics. He shared the responsibility in organizing several National Seminars on Social Science and Applied Economics Research. At present, he is working as a Research Professor and Director in the Department of Economic Policy Unit, Nigeria Institute of Social & Economics Research (NISER), Ibadan, Oyo State, Nigeria.

Dr. Olayiwola Olubodun Olaniyi was hail from Ilora community in Afijio Local Government of Oyo State, Nigeria. He completed his Ph.D in the year 2015 under the sponsorship of African Union Commission. Dr. Olayiwola Olubodun Olaniyi papers have been referred in various text and reference books published both national and international publishers.

Issues in Economic Development

Victor Adeyeye Abimbola

Olayiwola Olubodun Olaniyi

2017

Daya Publishing House®

A Division of

Astral International Pvt. Ltd.

New Delhi – 110 002

Cataloging in Publication Data--DK
Courtesy: D.K. Agencies (P) Ltd. <docinfo@dkagencies.com>

Adeyeye, V. A., author.
Issues in economic development / Victor Adeyeye Abimbola, Olayiwola Olubodun Olaniyi
pages cm
Includes bibliographical references and index.

ISBN 978-93-86071-51-4 (International Edition)

1. Economic development--Nigeria. 2. Nigeria--Economic policy. I. Title.

HC1055.Z9E26 2017 DDC 338.9669 23

Published by : **Daya Publishing House®**
A Division of
Astral International Pvt. Ltd.
– ISO 9001:2015 Certified Company –
4736/23, Ansari Road, Darya Ganj
New Delhi-110 002
Ph. 011-43549197, 23278134
E-mail: info@astralint.com
Website: www.astralint.com

Digitally Printed at : **Replika Press Pvt. Ltd.**

Foreword

The goal of the managers in the economy is always to close the gap between the potential and actual GDP, that is, remove the bottlenecks and create a situation for full employment and economy stability. It is an established fact that backward societies have not succeeded in making full use of their potentials for development not only because of the underdevelopment of their natural resources, including human resources and production technologies, but also due to the underdevelopment of the existing socioeconomic organization and institutions, including the network of markets and administrative machinery of the government in such societies. The general presumptions therefore, is that the elimination of these structural deficiencies in the economy of a backward society would enable such society, be it a nation, state or local government area, to make better use, not only of its domestics resources, but also its external economic opportunities in the form of international trade, and technological and organization innovations. It is in the pursuance of the foregoing that developing countries have always engaged in economic growth and development planning. The conventional plan attempts to promote economic growth and development in three main ways, namely:

(a) Increasing the overall rate of investment in the economy

(b) Engaging in some special type of investment designed to ease the bottlenecks of production in the important sector of the economy

(c) Trying to improve the coordination between different sectors and sub-sector of economy.

Lastly, the field of Agriculture economics would provide guidelines for making adjustment in the use of scares resources in a rational direction and prevent expending of resources in wrong direction and provide appropriate plans for government to execute their developmental policies and programmes.

Victor Adeyeye Abimbola

Preface

The book entitled *"Issues in Economic Development"* consists of twenty chapters. During the preparation of the manuscripts of the book, the author have incorporated the fruitful academic suggestions provided by senior staffs, colleagues and all the quoted statements inside the book are references. The authors are grateful to the entire staffs of Astral International (P) Ltd. for their interest in publishing the manuscript in its present shape. I additions, the constructive and academic suggestions received from all sides will always be welcomed and highly appreciated by the author.

Victor Adeyeye Abimbola

Olayiwola Olubodun Olaniyi

Contents

1
Key Issues in Economic Growth and Development Nexus in Nigeria

1.1. Key Issues in Economic Growth and Development Nexus in Nigeria

There is a global trend showing that increased growth rates have failed to answer questions like: what is happening to employment/unemployment? What is happening to the quality of education? What is happening to the provision of health services? What is happening to housing/shelter? What is the quality of governance, particularly, the participation of the citizens in how they are governed. "Growth without Development in Nigeria" Akpan H. Ekpo, TELL, Thursday, 29, March 2010. In Nigeria today many informed observers are debating the paradox of economic growth not accompanied by economic development. Recent available data from The National Bureau of Statistics (NBS) showed that the nation's Gross Domestic Product (GDP) is increasing at an impressive annual growth. Nigeria's GDP growth has been growing from 5.4% in 2005 to 7.8% in 2010 but fell to 6.7% in 2011. Growth is now driven by the non-oil sector. The slowdown in 2011 was a reflection of the worsening global economy and an oil production shut-down. For 2012, the economy is projected to grow by 6.9% on the back of stronger oil exports, but it will slow again in 2013 to 6.6%. (African Economic Outlook, 2011).

In several recent documents, (NEEDS, SAP) Nigeria acknowledged the developmental challenges facing the nation. Despite Nigeria's rich human and material resources endowment capable of making her Africa's largest economy and major player at the global level, yet much of the potential has remained untapped. Factors identified as hindering progress include: slow economic growth due to weak productive capacity, non- modernization of agriculture; low labour and total factor productivity, over dependency on crude oil production, lack of diversification of the economic base, gross inefficiency of public sector, poor and deteriorating social and economic infrastructural services, high dependence on primary production, heavy debt burden, low pace of technological advancement.

Through several reform programmes, initiated by current and past administrations, considerable and commendable progress has been achieved. Nonetheless, additional efforts are needed to confront the current growth without development scenario. This chapter hope to discuss

- The linkage of economic growth to economic development,
- Features of developing economy,
- Development planning and Nigeria's Economic Reform Programmes,
- Some key issues in Nigeria's growth and development nexus.

1.2. Economic Growth and Economic Development

Economic growth is a necessary but not sufficient condition of economic development. Economic growth and economic development are closely related but they are not synonymous. Economic growth can exist independent of the state of economic development. Economic development is essentially investment in an economy; economic growth is increased production of an economy. Economic growth is a term generally measured by the amount a country or region produces over a certain period of time. Economic growth is a narrower concept than economic development. It is an increase in a country's real level of national output which can be caused by an increase in the quality of resources, increase in the quantity of resources and improvements in technology or in another way an increase in the value of goods and services produced by every sector of the economy. Economic growth can be measured by an increase in a country's GDP (gross domestic product).

Economic development of a country is defined as the development of the economic wealth of the country. Economic development is aimed at the overall well-being of the citizens of country, as they are the ultimate beneficiaries of the development of the country. The definition of economic development given by Michael Todaro is an increase in living standards, improvement in self-esteem needs and freedom from oppression as well as a greater choice. The most accurate method of measuring development is the Human Development Index, HDI, which takes into account the literacy rates and life expectancy which affects productivity and could lead to economic growth. It also leads to the creation of more opportunities in the sectors of education, healthcare, employment and the conservation of the environment. Gallopin (1994), emphasized that development is not synonymous with economic growth. Development according to him involves qualitative transformation, while growth is essentially quantitative increase. The goal of development is to increase the quality of life of the human population; economic growth is only one mean to this end. Furthermore, quality of life embodies the satisfaction of material and non-material human needs and the fulfillment of human desires and aspiration.

Many writers argue that, after decades of the so called development project aimed at raising the Third World out of poverty and improving the well-being of its citizens, one can speak of anything but true transformation. Yet thousands of children still die of malnutrition-related diseases every day, and millions of people still do not have access to clean water. Inequalities within countries themselves persist. (Katarzyna Karpowicz, 2008). Development has long been a catch-all phrase and it now seems that everybody has something to say on the subject. It is indeed crucial to know what one is talking about, if only to allow for the differentiation of development from economic growth with which it is equated all too often. In point of fact, it is believed that the apparent failure of the development project can be attributed to precisely this misunderstanding surrounding the notion of development itself. In the words of Seabrook (1993: 7): "All over the world, more and more people are being disadvantaged by a version of development which, even if it creates wealth, leaves them with a sense of loss and impoverishment". Thus, based largely on the comparison of two indicators used by major international bodies dealing with development – Gross Domestic Product (GDP) and Human Development Index (HDI), it is argued that growth and development are, actually, two different phenomena. In fact, economic growth is but a step in the direction towards development – one of major significance, indeed a precondition to it, but by no means can it be conceptualized as development itself. For a country to be generally recognized as a developed one, it also needs to be able to provide its citizens with basic resources and services, such as food, healthcare and schooling.

The problem with seeing development purely in terms of a country's economic performance is clearly visible if one considers the method used for measuring it. For a long time, the indicator used for measuring development was Gross Domestic Product (GDP) expressed in per capita terms. Those who support this view of development would have it that unless a rise in GDP per capita takes place, no development whatsoever is bound to happen (Storey 2003: 29). What follows is that the increase in GDP per capita is a sine qua non of development. True as it may be the fact that GDP is a purely economic indicator and, as such, does not allow for a change in social, cultural or political conditions. Also, it has been blamed for its falling short of providing a true image of the country's development, or the lack of it. Another major problem with GDP, and thus with the perception of economic growth as equal to development, is that it gives no indication of how fairly resources are distributed in the country. If development cannot solely be based on economic growth, as reflected in the increase in GDP, then what else does it encompass? Having stated that economic growth is but one of many elements of development, one may then try to look at some of the other factors that would be included in the definitions of development. In doing so, we may want to consider the UN's Human Development Index (HDI). HDI has been designed to remedy the shortcomings of GDP as a measurement of the progress achieved by developing countries. It not only allows for the growth in income per capita, but comprises also two other elements crucial to human well-being: life expectancy and educational attainment. Thus, development is seen as a "process of enlarging people's choices" in order for the people to "lead long and healthy lives, to be knowledgeable and to have a decent standard of living" (Human Development Reports.)

Therefore, HDI is a comparative measure of life expectancy, literacy, education and quality of life. It is a standard of measuring well-being, and is used to distinguish whether a country is a developed, developing or an underdeveloped country. HDI was introduced as an alternative to conventional measure of national development such as level of income and rate of economic growth. Nigeria HDI is 0.459, which gave the country a rank of 156 out of 187 countries with comparable data.

1.3. Features of Developing Economy

We may now identify some features of an underdeveloped economy from that of developed one as follows:

1. In an underdeveloped economy, the fundamental problem is that of output, real income or the standard of living, as these economies are characterized by low productivity, low income and a poor standard of living. A vast majority are undernourished and without adequate shelter. On the other hand, most of the developed countries at present enjoy a high rate of mass-consumption. In this economies, per capital real income has risen to a level at which a large number of people can afford consumption transcending food, shelter and clothing.
2. An underdeveloped economy, compared with an advanced economy, is underequipped with capital in relation to its population and natural resources. The rate of growth of employment and investment in such an economy lags behind the rate of growth of population. The resources are not only employed but also underemployed.
3. High rate of growth of population is an important characteristic of most of the underdeveloped economies. Population growth in underdeveloped countries neutralizes economic growth.
4. The central problem of underdeveloped economies is the prevalence of mass poverty which is the cause as well as the consequence of their low level of development. Shortage and scarcity are the main economic problems in these economies, whereas the affluent societies of advanced countries have economic problems resulting from abundance.
5. There is chronic unemployment in an underdeveloped economy. An advanced economy may have unemployment occasionally due to business fluctuations and a low marginal propensity to consume. Whereas an under developed economy is confronted with the problem of disguised unemployment in the sense that even with unchanged techniques in agriculture could be removed without reducing agricultural output.
6. Poor countries are poor in technology; advanced countries are advanced in technology. In fact, the level of technology attained in production is a reliable indication of the level of economic development. Employment

of advanced technology goes along with large capital resources, high attainments in the fields of scientific research, greater availability of entrepreneurial skill and a good supply of efficient skilled labour. Thus, development of technology is the basic objective of the backward economy.

1.4. Development Planning and Nigeria's Economic Reform Programmes

(a) National Development Plans (NDP)

After independence in 1960 the need for a more formal and coordinated economic development resulted in the 1962-68 plan, which represented the first "national" development plan because of the extensive involvement of Nigerians in its formulation. One notable characteristic of the plan was the extent of coordination between the central government and the federating units. The Second National Development Plan, (1970-74), was more comprehensive and ambitious as crude oil production had already emerged as a major contributor to national income. The major focus of the 1970-74 Plan was the reconstruction of the war-torn economy. The objectives espoused were to establish a united, strong and self-reliant nation, develop a great and dynamic economy, provide the basis for a just and egalitarian society, transform Nigeria into a land of bright and full opportunities for all citizens and develop a free and democratic society.

Subsequently, two more development plans were initiated *i.e.* The Third National Development Plan (1975-80), and the Fourth National Development Plan (1981-85). The third plan involved extensive consultation which the private sector and the academia, and because Nigeria witnessed unprecedented oil boom in the mid 1970s, it benefited from the increased earnings from oil production. The main objectives did not diverge much from that of the earlier plan, though there was an added objective, namely, "indigenization of economic activities". The 1981-85 Plan, which succeeded it, was not remarkably different in objectives, except that it was introduced under a democratic administration. From the early part of the fourth plan period, Nigeria witnessed a lot of economic problems owing to excessive external reserve loss arising from capital flight, worsening terms of trade and drastic fall in domestic production. The economy was not performing at its optimal level. Rationing and price control were rampant, and as it the case with planned economies, there was always a shortage of some items and a surplus of others. In the midst of the problem, the plan was virtually abandoned. On top of the economic problems was a major political crisis, which culminated in the overthrow of the civilian government. The new military government promised accountability and was able to achieve some level of internal and external balance by reducing public spending and foreign exchange commitments through further import restrictions

(b) Structural Adjustment Programmes (SAP)

The emergency of economic crises in the Nigeria society was the main cause of instituting the structural adjustment programme (SAP). To avert or moderate the effects of the economic crises, the government towards the middle of 1986 announced the commencement of SAP for the period July, 1986 to June 1988. The aim is to effectively alter and restructure the consumption and production patterns of the economy as well as eliminate price distortions and dependence on oil revenue and import of consumer and producer goods which then characterize the economy. In addition, SAP is expected to enhance the non-oil export, especially the tradable agricultural products thereby improving the economy and bringing the economy back on the path of steady and balance growth. The main strategies of SAP include: institutional reform; pricing policy and specific production scheme for agricultural crops. Economic reform in Nigeria in the 1990s witnessed a review of SAP because of its limited impact.

(c) Review of National Economic Empowerment and Development Strategy (NEEDS)

However, the advent of democratic rule in 1999 brought about a rethinking of economic reform in Nigeria. The widespread disillusionment arising from the failure of previous reforms and the experiences gained from the participatory approaches employed in the development of the interim Poverty Reduction Strategic Paper (PRSP) are the catalysts of a new social and economic development framework termed "National Economic Empowerment and Development Strategy (NEEDS). NEEDS provided the basis and philosophical undertone for economic reform. It has four main strategic features that determine the nature and scope of economic reform measures:

- Reforming the way government works and its institutions;
- Growing the private sector;
- Implementing a social charter for the people; and
- Re-orientation of the people with an enduring African value system.

The NEED document involved a participatory process, with the State and local government levels keying into the programme through their own versions- the State Economic Empowerment and Development Strategy (SEEDS) and the Local Economic Empowerment and Development Strategy (LEEDS). These comprised broad agendas of social and economic reforms based on four key strategies, including private sector participation as the main driver of the economy; implementation of a social charter for improved security and welfare; value re-orientation; and focus on attracting Foreign Direct Investment (FDI) (OSGF, 2012).

(d)Vision 20:2020

In 2007, Vision 20:2020 was unveiled by the government, declaring the intention of catapulting Nigeria into the group of twenty top economies by the year 2020. The vision was anchored on the Seven Point Agenda (SPA) of Yar'Adua/Goodluck administration, with the objectives of meeting the energy, transportation, human capital development, land tenure, food security needs of Nigerians, among others.

(e) Transformation Agenda

There is indeed a great concern about the failure of previous strategies to achieve the desired effects on the development of Nigeria. Probable ineffectiveness in policy implementation inevitably caused the failure of previous efforts at repositioning Nigeria. The emergence of President Goodluck Jonathan in 2010/2011 set the background for a new order–a search for a strategy for the recreation of Nigeria. This, undoubtedly, necessitated the introduction of the Transformation Agenda (OSGF, 2012). According to the Office of the Secretary to the Government of the Federation (OSGF) (2012), the cardinal elements of the Transformation Agenda of the current administration include constitutional and electoral reforms and transformation of the budgeting process. Others will include overcoming the current security challenges through ensuring peaceful co-existence, tolerance, employment generation, poverty alleviation and job creation for our youths aside from determined focus on developing critical infrastructure and human capital development. There is an overriding push for effective planning, coordination and continuity of policies and programmes with creation of an enabling environment for private sector participation through Public Private Partnership (PPP). The Administration is also poised to deliver good governance through tackling corruption, value re-orientation, accountability and transparency as well as Monitoring and Evaluation for all priority policies, programmes and projects.

2

Key Issues in Growth and Development in Nigeria

2.1. Key Issues in Growth and Development in Nigeria

Nigeria has attempted through various reform programmes to better the lots of her citizens by providing dividend of development, such as education, health, social infrastructure. Broadly speaking, the features of an underdeveloped economy create obstacles in the way of economic development, and hamper economic progress. These features emerge out of economic, social, political, religious and institutional factors. It would be wrong however to conclude that only economic factors are responsible for poverty or economic backwardness of a country. Non-economic factors are equally responsible for the under development of an economy. The factors discouraging economic development may be classified into economic and non-economic factors which are:

Economic Factors	*Non-Economic Factors*
1. Vicious Circle of Poverty	1. Underdeveloped Human Resources
2. Deficiency of Capital	2. Political Instability
3. Market Imperfections	3. Socio-cultural constrains
4. International Forces	4. Religious factors
5. Low Agricultural Productivity	
6. Lack of Entrepreneurs	

Available evidence from other countries seems to suggest to some extent that economic growth lead to economic development when properly administered. In Nigerian today it is apt to ask: **Why is economic growth not leading to development?** With respect to the current situation in Nigeria, I will briefly discuss the following issues relevant to development among several others.

2.2. Corruption and Governance Challenge in Nigeria

The most basic requirement for faster rate of development is improved governance, stability and peace. Some observers have noted that absence of these virtues is inimical to sustainable development. Nigeria is currently being affected by these problems to the extent that in large areas of the nation real development is impeded. The causes/reasons for this situation varied across the nation but the underlining causes could be traced to poverty, under-development, and lack of economic diversification and marginalization of the population. Unless these issues are redress, development however perceived will be difficult to achieve despite high GDP. Corruption is a major obstacle to meaningful development in Nigeria. This fact is acknowledged locally and internationally. Government has however put in place institutions to combat this menace, but we all know that the problem is far from over.

Recently, the World Bank in its report, as reported in Punch Newspaper (August, 10, 2012) noted that" investors in Nigeria lost 10% of their revenue to poor quality infrastructure, crime, insecurity and corruption in 2011". The 2011 Corruption Perception Index (CPI) published by Transparency International ranked Nigeria as 143 out of 183 surveyed countries. The CPI for Nigeria is 2.2 (with 10.0 being a clean country and 0 being the most corrupt). The government and in particular, the people of Nigeria must agree to put a stop to corrupt practices as government alone is not capable without the cooperation of the populace. The current report of foreign firms relocating to nearby nations is a clear reflection of the negative impact of corruption and insecurity on development.

In a remark on Institutional and Governance Reforms in Nigeria by the The Brookings Institution (2007), it noted with concern the effect of corruption on the growth and development of the nation. The report posited thus: "a bane of Nigeria's existence since the oil boom of the 1970s has been the reputation for corruption, largely justified, but also partly the result of perception. Corruption and poor governance affected growth and public service delivery in Nigeria in various ways. Corruption distorts the climate for doing business and serves as a tax on private investments. In a corrupt environment, resources for human capital and other needed investments, such as in infrastructure, health and education are often diverted. There are various ways in which this may occur, including procurement fraud, patronage for access to services, absenteeism and misuse of facilities. In particular, poor households are disproportionately hurt in communities where corruption is most prevalent. A key tenet of the reform program, therefore, was to prioritize anti-corruption measures as a central component of the reforms."

Effort at combating corruption in Nigeria is an on-going but complex problem. Corruption has devastating effect on the nation, reduces our citizens to poverty and deprived the nation of any sustainable development. As a result of massive corruption, Nigeria faces many challenges in sustaining economic growth and improving its broad development indicators.

2.3. Addressing Poverty and Inequality

The current Nigerian Poverty Profile Report of the Nigerian Bureau of Statistic 2010 (NBS) reveals a disappointing scenario. The following measurements are used to evaluate poverty in Nigeria:

a) Relative Poverty (RP) is defined by reference to the living standard of majority of a given society. In 2004, the Nigerian RP measurement was 54.4% but increased to 69.0%, or 112.518million, in 2010.

b) Absolute Poverty (AP) is defined in term of the minimal requirement necessary to afford minimal standard of food, shelter, clothing etc. 54.7% Nigerians were living in poverty in 2004 but this increased to 60.9% or 99.28 million Nigerians.

c) The Dollar-per-day measurement refers to the proportion of those living on less than $1 per day poverty line. With respect to this measurememt,51.6% Nigerians were living below US$1 per day in 2004 but increased to 61.2%

Income inequality, measured by Gini coefficient rose from 0.429 in 2004 to 0.447 in 2010 indicating greater income inequality during the period. Forecast by NBS for 2011 is not cheering as it estimate that poverty, using relative, absolute and dollar-per-day may further increase to about 71.5%, 61.9% and 62.8% respectively. It is an unexplained paradox, a self –contradictory statement, that despite the fact that Nigerian economy is said to be growing, the proportion of people living in poverty increases yearly, because economic growth is not development and economic growth has not been successfully translated to development through appropriate government policies. Consequently, he number of poor people continues to increase despite several development and poverty reduction programmes and rural citizens constitute the majority of the population and poverty is deemed to be acute in the rural relative to the urban sectors. The number of poor people will continue to increase except correct national steps are put in place involving the opinions and consultation with poor rural dwellers. In effect, poverty assessment and reduction must be participatory.

Nigeria is a large country with higher concentration of rural dwellers. The rural dwellers are about 80% of the population. Economic development in Nigeria is unlikely to succeed unless the issue of (rural) poverty is addressed. The problems associated with poverty are enormous. They include issues such as: education, health, land tenure, credit, production and marketing processes.

The causes of poverty in Nigeria are very complex but the basic causes are closely related to problems of access and endowment. Specifically, they are according to the World Bank (1985);

- Inadequate access to employment opportunity for the poor;
- Inadequate asset such as land and capital for the poor;

- ☆ Inadequate access to the means of authority in poor regions;
- ☆ Inadequate access to market for goods and services which the poor can sell;
- ☆ Inadequate access to education, health, and water services;
- ☆ Inadequate involvement of the poor in the design of development programmes;

Regarding the characteristics of poverty, it can be safely assumed that:

(i) While rural poverty dominates in Nigeria, urban poverty is a growing and explosive problem propelled by employment crises especially of the youth.

(ii) Gender inequalities are a key feature of poverty pattern.

The character of poverty as described above seems to suggest that the poverty reduction strategy will include:

(a) Labour intensive growth;

(b) Agricultural growth and provision of basic health services, education and infrastructure in particular to rural areas;

(c) Investment in human resources, encouragement of growth of investment and employment in urban areas with the focus on the informal sector.

2.4. Building Human Resources

Building human resources involve investing in people, through education and health. Education, including skills acquisition is the basic foundation for development. A literate and skilled society is better organized, productive, active and development oriented. Also investing in people is essential for growth, development and poverty reduction. Developed countries have invested massively and continuously on human resources development. Given the present situation, Nigeria needs to invest on its human resources development through adequate education of its youth, via skill acquisition and formal education of its people, especially women. Goal 3 of the MDG is to Promote Gender Equality and Empower Women. Recent evaluation however shows that a gradual improvement in the proportion of girls enrolled in primary school, though noteworthy, is not yet enough to meet the target. There are still fewer girls than boys in school. There are signs of backsliding in the number of girls in tertiary education. Measures to encourage girls to attend school, particularly by addressing cultural barriers in the north of the country, and to provide the economic incentives for boys to attend school in the south-east, are urgently required. Confronting regional variations in the determinants of gender inequality requires policies based on an understanding of the underlying socioeconomic, social and cultural factors. State and local government efforts will thus be critical to the achievement of this goal (UNDP, 2007).

The government has been implementing education reforms since 2006 to improve access to education, especially with the introduction of Universal Basic Education. The government increased the share of education in budget spending from 4% in 2010 to 6% in 2011. There have been some positive results. The literacy rate for the 15-24 age groups was 80% in 2008, up from 64.1% in 2000. Education quality is, however, generally low and varies considerably across the country. The West African Examination Council said that in 2010 only 25% of Nigerian candidates obtained credits in English language, mathematics, and at least three other subjects (the minimum requirement to enter a tertiary institution). Social indicators on health and education remain weak. The 2011 UN Human Development Report ranked Nigeria 156th out of 187 countries. In 2010, Nigeria was 142nd out of 169 countries. The government has sought to improve the health care system. Health spending in the federal budget increased from 4% in 2010 to 6% in 2011. Also, the Federal Government has adopted several policy initiatives to strengthen the health system, including a National Strategic Health Development (NSHD) plan. Despite these efforts, health outcomes remain poor. According to the United Nations 2010 MDG Report, the infant mortality rate was 75 deaths per 1 000 live births, the under-five mortality rate was 157 per 1 000, the maternal mortality rate was 545 deaths per 100 000 live births and the proportion of births attended by skilled personnel was 39%.

2.5. Making Agricultural Development: A Priority in Nigeria

Despite oil's revenue dominance, agriculture plays a significant role in the Nigerian economy, accounting for 34.5% of GDP in the first quarter of 2012 according to the National Bureau of Statistics. Sustainable expansion of agriculture should play a key role in unleashing inclusive economic growth, reducing poverty and enhancing food security. The government's Agricultural Transformation Agenda targets these objectives. Agriculture's importance to poverty reduction goes far beyond its direct impact on farmers' incomes. There is evidence that increasing agricultural productivity has benefited millions through higher incomes, more plentiful and cheaper food, and by generating patterns of development that are employment-intensive and benefit both rural and urban areas. More importantly, it has provided the spur to economic development outside agriculture where growth and job creation are faster and wages higher. Rapid increases in agricultural output, brought about by increasing land and labour productivity, have made food cheaper, benefiting both the urban and rural poor, who spend much of their income on food.

In a special report on Development Outreach (vol.10, no.3, 2008), the editor advocated for making agriculture a priority and called for renewed effort to place agriculture at the center of the development agenda in order to meet the MDG goal of halving extreme poverty and hunger by 2015. As an economic activity agriculture contributes to development as a source of livelihood and food security and provider of environmental services. Agriculture can be:

- ✰ The main source of growth for national economy,
- ✰ Provider of investment opportunities for the private sector,
- ✰ A private driver of agric-related industries and rural non-farm economy

Majority of poor people live in the rural area and they live on less than $2 a day and must depend on agriculture. Given where they live and what they do, promoting agriculture is imperative for meeting the MDG of halving poverty and hunger by 2015. And continue to reduce poverty and hunger several decades after. Though agriculture alone will not be enough to massively reduce poverty, but poverty reduction will not happen without agriculture.

Poverty is overwhelmingly a rural problem. That is so, and that it endures is the failure of agriculture to deliver livelihood for the large proportion of our people who depend on it. Agriculture growth can reduce poverty directly, by raising farm income and indirectly, through labour market and by reducing food prices. Increased farm productivity increase farm income and participation of smallholders in this agricultural growth will obviously benefit them. Agricultural growth also reduces poverty via labor market because agricultural growth is capable of creating employment opportunities for the poor. Lastly, ability to increase non-tradable staple foods will reduce domestic food prices, especially to the poor, in addition to the urban dwellers, who spend a substantial amount of their income on food.

Agriculture has a special poverty reducing power. China, India and Ghana, for example, have all had relatively high agricultural growth rates and reduced poverty substantially as a result. As noted by Development Outreach (vol. 10, no. 3, 2008), Ghana's growth and poverty reduction over the last 15 years is Africa's success story. Real GDP has grown at more than four per cent a year since 1980 and at more than five per cent since 2001. The poverty rate fell from 52% in 1991 to 29% in 2005, making Ghana the only country in sub- Saharan Africa to have already met the MDG goal of halving poverty.

Similarly, IFAD (2005) in its publication noted "that reversing recent disappointing trends in agriculture's performance is critical if poor countries are to escape the trap of slow growth and poverty. This is particularly true in sub-Saharan Africa, where growth in agricultural output has barely kept pace with population. Productivity has stagnated, slowing wider economic growth and exacerbating poverty with it."

2.6. Conclusion

Economic development refers to a sustainable increase in living standard. It implies higher per capita income, better education and health and environmental protection. Public policy must generally aim at continuous national and sustained economic growth and expansion of the nation's economy so that "developing countries" become "developed". The economic development process involves the adjustment of national institutions to give incentives for innovation and investment so as to develop an efficient production and distribution system of goods and services. When discussing the link of growth to development, one often hears that whereas growth is quantity, development is quality. Simplistic as this comparison is, it shows the inherent difference between the two phenomena. Indeed, the actual income of a country is of relatively lesser importance when compared to the way in which this relative wealth translates into the quality of services the state renders to its citizens. And it is precisely those services – education, healthcare, the provision of basic resources such as food or water – that define how developed a country really is. It is essential that policy-makers and all those preoccupied with development, professionally or otherwise, are fully aware of this seminal difference. Until such holistic definition of development is not only adopted on an international arena, but also truly applied in all actions related to the countries of the so called Third World, the gap between the rich and the poor will keep widening, and the answer to the question: "Is development a failure?", will always be positive. (Katarzyna Karpowicz, 2008). Economic development is a sustainable boost in the standard of living of the people of a country. It implies an increase in the per capita income of every citizen. It also leads to the creation of more opportunities in the sectors of education, health, employment and the conservation of the environment.

3
Comparative Economic System

3.1. Comparative Economic System

The comparative economic system can be describe as hoe a nation is organised for production of goods and services weather it has a traditional market or a command economy. The comparative economics is also an alternative way of solving or answering the three economics questions or problems *i.e.,* how will goods be produced, what will be produced, for whom will goods will be produced? All economic system, regardless of the decision making process must answer these three fundamental questions. Nations answer these questions differently depending on who owns their resources and how their economic activities are conducted and directed. Two extreme ownership and coordination are represented by pure capital and pure command economy. All countries exemplify a mix of both element with some learning more towards capitalism and others learning towards a command economy.

An economic system is comprised of the various processes of organizing and motivating labour, producing, distributing, and circulating of the fruits of human labour, including products and services, consumer goods, machines, tools, and other technology used as inputs to future production, and the infrastructure within and through which production, distribution, and circulation occurs. These processes are over determined by the political, cultural, and environmental conditions within which they come to exist. In comparative economic systems, these economic systems are usually defined within determinate political boundaries. Thus, one would speak of a Chinese economic system, although China may, in fact, be a complex conglomeration and interaction of economic systems. Nevertheless, bounding economic systems in this way provides a way of discussing how such systems are made possible and changed by the specific effects of politico-institutional, cultural, and environmental differences. Thus, one might discuss how the capitalist economic system of 1999 Germany is different from the capitalist economic system of 1999 Britain, for example

A mainstream comparative economic system was, for a long time, largely a propaganda tool used in the Cold War. Thus, there was very little science employed. Instead, the course was used to promote the NATO economies, primarily the United States and Great Britain, as the best possible economic systems and the CMEA economies, primarily the USSR, as the worst possible economic systems. Systemic similarities between economies across this NATO/CMEA divide were ignored, as were careful definitions of economic, political, and cultural terms. The possibility that Stalinism was compatible with a form of capitalism, for example, was never explored, despite numerous examples of totalitarian political systems wedded to relatively open (free-market) versions of capitalism. More recent mainstream approaches to comparative economic systems are less polemical, focusing on the way non-economic factors influence (as exogenous variables) the structure and functioning of economies.

An economic system is comprised of the various processes of organizing and motivating labor, producing, distributing, and circulating of the fruits of human labor, including products and services, consumer goods, machines, tools, and other technology used as inputs to future production, and the infrastructure within and through which production, distribution, and circulation occurs. These processes are over determined by the political, cultural, and environmental conditions within which they come to exist. In comparative economic systems, these economic systems are usually defined within determinate political boundaries. Thus, one would speak of a Chinese economic system, although China may, in fact, be a complex conglomeration and interaction of economic systems. Nevertheless, bounding economic systems in this way provides a way of discussing how such systems are made possible and changed by the specific effects of politico-institutional, cultural, and environmental differences. Thus, one might discuss how the capitalist economic system of 1999 Germany is different from the capitalist economic system of 1999 Britain, for example Mainstream comparative economic systems was, for a long time, largely a propaganda tool used in the Cold War. Thus, there was very little science employed. Instead, the course was used to promote the NATO economies, primarily the United States and Great Britain, as the best possible economic systems and the CMEA economies, primarily the USSR, as the worst possible economic systems. Systemic similarities between economies across this NATO/CMEA divide were ignored, as were careful definitions of economic, political, and cultural terms. The possibility that Stalinism was compatible with a form of capitalism, for example, was never explored, despite numerous examples of totalitarian political systems wedded to relatively open (free-market) versions of capitalism. More recent mainstream approaches to comparative economic systems are less polemical, focusing on the way non-economic factors influence (as exogenous variables) the structure and functioning of economies.

3.2. Classification

1. Centrally planned or command economy
2. Free market economy
3. Traditional or mixed economy

3.2.1. Centrally Planned or Command Economy

This is an economy where all economic decisions are taken by the central authorities or Government. The main decision maker is the Government; no person may independently decide t open and run a business. The Government decide what goods and services are to be produced and owns the places where the production takes place. The Government sells this goods and services and decides how the talents and skills of its workers are to be used. In theory, the command economy incorporates individual choice into collective choices which in turn is reflected in central planning decision *e.g.* China.

3.2.2. Free Market Economy of Capitalism

A market or capitalism economic system is one in which a economic decisions are the result of individual decisions by buyers and seller in the market. Market systems are also known as "free enterprise system". Capitalism is an economic system characterised by private ownership of resources and the use of prices to coordinates economic activities in free, competitive markets. Any income derived from the use of land, labour, capital, entrepreneurial ability goes to the individuals owners of those resources. Owners have property rights to the use of their resources and are therefore free to sell their resources to the highest bidder. Consumers are free to buy whatever goods they can afford while producers are free to make and sell whatever output they think will be profitable *e.g.* US, North Korea, Cuba.

3.2.3. Mixed Economic System

Because of the problem of both free market and command economics, all real –world economics are a mixture of the two systems. In the mixed economy the decision is taken by consumers and businesses and by government determines economic activities. Hungary and Poland are also experimenting with a professional of small, consumer oriented businesses. Finally some economic system are directed largely by custom or religion. Laws of the Muslim religion set limits and the rate or interest that can be earned on certain investments. The caste system in India and elsewhere often restricts the occupations available to individuals hence, religion, custom and family relations play important roles in the organisation and co ordinations of economic activity. Your own pattern of consumption and choice of occupation may be influenced by some of these factors.

3.3. Nigeria Economic System

Is a middle income, mixed economy emerging market with well-developed financial, legal, communications, transport, and entertainment sectors. It is ranked 31st in the world in terms of GDP (PPP) as of 2009, and its emergent, though currently underperforming manufacturing sector is the second-largest on the continent, producing a large proportion of goods and services for the West African region. Previously hindered by years of mismanagement, economic reforms of the past decade have put Nigeria back on track towards achieving its full economic potential. Nigerian GDP at purchasing power parity more than doubled from \$170.7 billion in 2005 to \$374.3 billion in 2010, although estimates of the size of the informal sector (which is not included in official figures) put the actual numbers closer to \$520 billion. Correspondingly, the GDP per capita doubled from \$1200 per person in 2005 to an estimated \$2,500 per person in 2009 (again, with the inclusion of the informal sector, it is estimated that GDP per capita hovers around \$3,500 per person). It is the largest economy in the West Africa Region, 3rd largest economy in Africa (behind South Africa and Egypt), and on track to becoming one of the top 30 economies in the world in the early part of 2011.

Although much has been made of its status as a major exporter of oil, Nigeria produces only about 3.3% of the world's supply, and though it is ranked as 15th in production at 2.2 million barrels per day (350,000 m^3/d) (mbpd), the top 3 producers Saudi Arabia, Russia, and the United States produce 10.7 Mbbl/d (1,700,000 m^3/d) (17%), 9.8 Mbbl/d (1,560,000 m^3/d) (15%), and 8.5 Mbbl/d (1,350,000 m^3/d) (13%) respectively, collectively accounting for 63.6 Mbbl/d (10,110,000 m^3/d) (45%) of the world's total production. [4] To put oil revenues in perspective: at an estimated export rate of 1.9 Mbbl/d (300,000 m^3/d), with a projected sales price of \$65 per barrel in 2011, Nigeria's anticipated revenue from petroleum is about \$52.2 billion. This accounts for less than 14% of official GDP figures (and drops to 10% when the informal economy is included in these calculations). Therefore, though the petroleum sector is important, it remains in fact a small part of the country's overall vibrant and diversified economy. The largely subsistence agricultural sector has not kept up with rapid population growth, and Nigeria, once a large net exporter of food, now imports some of its food products. In 2006, Nigeria successfully convinced the Paris Club to let it buy back the bulk of its debts owed to the Paris Club for a cash payment of roughly \$12 billion (USD).

Nigeria's economy is struggling to leverage the country's vast wealth in fossil fuels in order to displace the crushing poverty that affects about 57 per cent of its population. Economists refer to the coexistence of vast wealth in natural resources and extreme personal poverty in developing countries like Nigeria as the "resource curse". Although "resource curse" is more widely understood to mean an abundance of natural resources which fuels official corruption resulting in a violent competition for the resource by the citizens of the nation. Nigeria's exports of oil and natural gas–at a time of peak prices–have enabled the country to post merchandise trade and current account surpluses in recent years. Reportedly, 80 per cent of Nigeria's energy revenues flow to the government, 16 per cent cover operational costs, and the remaining 4 per cent go to investors. However, the World Bank has estimated that as a result of corruption 80 per cent of energy revenues benefit only 1 per cent of the population. In 2005, Nigeria achieved a milestone agreement with the Paris Club of lending nations to eliminate all of its bilateral external debt. Under the agreement, the lenders will forgive most of the debt, and Nigeria will pay off the remainder with a portion of its energy revenues. Outside of the energy sector, Nigeria's economy is highly inefficient. Moreover, human capital is underdeveloped–Nigeria ranked 151 out of 177 countries in the United Nations Development Index in 2004–and non-energy-related infrastructure is inadequate. From 2003 to 2007, Nigeria attempted to implement an economic reform program called the National Economic Empowerment Development Strategy (NEEDS). The purpose of the NEEDS was to raise the country's standard of living through a variety of reforms, including macroeconomic stability, deregulation, liberalization, privatization, transparency, and accountability. The NEEDS addressed basic deficiencies, such as the lack of freshwater for household use and irrigation, unreliable power supplies, decaying infrastructure, impediments to private enterprise, and corruption. The government hoped that the NEEDS would create 7 million new jobs, diversify the economy, boost non-energy exports, increase industrial capacity utilization, and improve agricultural productivity. A related initiative on the state level is the State Economic Empowerment Development Strategy (SEEDS).

A longer-term economic development program is the United Nations (UN)-sponsored National Millennium Goals for Nigeria. Under the program, which covers the years from 2000 to 2015, Nigeria is committed to achieve a wide range of ambitious objectives involving poverty reduction, education, gender equality, health, the environment, and international development cooperation. In an update released in 2004, the UN found that Nigeria was making progress toward achieving several goals but was falling short on others. Specifically, Nigeria had advanced efforts to provide universal primary education, protect the environment, and develop a global development partnership. However, the country lagged behind on the goals of eliminating extreme poverty and hunger, reducing child and maternal mortality, and combating diseases such as human immunodeficiency virus/acquired immune deficiency syndrome (HIV/AIDS) and malaria.

A prerequisite for achieving many of these worthwhile objectives is curtailing endemic corruption, which stymies development and taints Nigeria's business environment. President Olusegun Obasanjo's campaign against corruption, which includes the arrest of officials accused of misdeeds and recovering stolen funds, has won praise from the World Bank. In September 2005, Nigeria, with the assistance of the World Bank, began to recover US$458 million of illicit funds that had been deposited in Swiss banks by the late military dictator Sani Abacha, who ruled Nigeria from 1993 to 1998. However, while broad-based progress has been slow, these efforts have begun to become evident in international surveys of corruption. In fact, Nigeria's ranking has consistently improved since 2001 ranking 147 out of 180 countries in Transparency International's 2007 Corruption Perceptions.

Table 3.1: Monetary Fund with Figures in Millions of Nigerian Naira

Year	*Gross Domestic Product*	*US Dollar Exchange*	*Inflation Index (2000=100)*	*Per Capita Income (as % of USA)*
1980	50,849	0.78 Naira	1.30	7.22
1985	98,619	2.83 Naira	3.20	1.87
1990	286,374	8.94 Naira	8.10	1.49
1995	1,928,642	54.36 Naira	56	1.28
2000	4,676,394	102.24 Naira	100	1.11
2005	14,894,454	131.01 Naira	207	1.96
2010	25,647,201	160.04 Naira	406	2.42
2014	31,058,452	196.48 Naira	482	2.72

Source: CBN (2015).

For purchasing power parity comparisons, the US Dollar is exchanged at 75.75 Nigerian Naira only. Current GDP per capita] of Nigeria expanded 132% in the Sixties reaching a peak growth of 283% in the Seventies. But this proved unsustainable and it consequently shrank by 66% in the Eighties. In the Nineties, diversification initiatives finally took effect and decadal growth was restored to 10%. Due to inflation, per capita GDP today remains lower than in 1960 when Nigeria declared independence. About 57 per cent of the population lives on less than US$1 per day. In 2005 the GDP was composed of the following sectors: agriculture, 26.8 per cent; industry, 48.8 per cent; and services, 24.4 per cent. In 2005 Nigeria's inflation rate was an estimated 15.6 per cent. Nigeria's goal under the National Economic Empowerment Development Strategy (NEEDS) programme is to reduce inflation to the single digits. In 2005 Nigeria's central government had expenditures of US$13.54 billion but revenues of only US$12.86 billion, resulting in a budget deficit of 5 per cent. Nigerian tax authorities face the challenge of widespread tax evasion, which is motivated by complaints about corruption and the poor quality of services.

Mean wages were $1.11 per man-hour in 2013.

(a) Agriculture

Nigeria ranks twenty fifth worldwide and first in Africa in farm output.

Agriculture has suffered from years of mismanagement, inconsistent and poorly conceived government policies, and the lack of basic infrastructure. Still, the sector accounts for over 26.8% of GDP and two-thirds of employment. Nigeria is no longer a major exporter of cocoa, groundnuts (peanuts), rubber, and palm oil. Cocoa production, mostly from obsolete varieties and overage trees, is stagnant at around 180,000 tons annually; 25 years ago it was 300,000 tons. An even more dramatic decline in groundnut and palm oil production also has taken place. Once the biggest poultry producer in Africa, corporate poultry output has been slashed from 40 million birds annually to about 18 million. Import constraints limit the availability of many agricultural and food processing inputs for poultry and other sectors. Fisheries are poorly managed. Most critical for the country's future, Nigeria's land tenure system does not encourage long-term investment in technology or modern production methods and does not inspire the availability of rural credit.

Agricultural products include cassava (tapioca), corn, cocoa, millet, palm oil, peanuts, rice, rubber, sorghum, and yams. In 2003 livestock production, in order of metric tonnage, featured eggs, milk, beef and veal, poultry, and pork, respectively. In the same year, the total fishing catch was 505.8 metric tons. Roundwood removals totaled slightly less than 70 million cubic meters, and sawnwood production was estimated at 2 million cubic meters. The agricultural sector suffers from extremely low productivity, reflecting reliance on antiquated methods. Although overall agricultural production rose by 28 per cent during the 1990s, per capita output rose by only 8.5 per cent during the same decade. Agriculture has failed to keep pace with Nigeria's rapid population growth, so that the country, which once exported food, now relies on imports to sustain itself.

(b) Industry

Nigeria ranks 44th worldwide and third in Africa in factory output.

The oil boom of the 1970s led Nigeria to neglect its strong agricultural and light manufacturing bases in favor of an unhealthy dependence on crude oil. In 2000, oil and gas exports accounted for more than 98% of export earnings and about 83% of federal government revenue. New oil wealth, the concurrent decline of other economic sectors, and a lurch toward a statist economic model fuelled massive migration to the cities and led to increasingly widespread poverty, especially in rural areas. A collapse of basic infrastructure and social services since the early 1980s accompanied this trend. By 2000, Nigeria's per capita income had plunged to about one-quarter of its mid-1970s high, below the level at independence. Along with the endemic malaise of Nigeria's non-oil sectors, the economy continues to witness massive growth of "informal sector" economic activities, estimated by some to be as high as 75% of the total economy.

Nigeria's proven oil reserves are estimated to be 35 billion barrels (5.6×10^9 m^3); natural gas reserves are well over 100 trillion cubic feet (2,800 km^3). Nigeria is a member of the Organization of Petroleum Exporting Countries (OPEC), and in mid-2001, its crude oil production was averaging around The types of crude oil exported by Nigeria are Bonny light oil, Forcados crude oil, Qua Ibo crude oil and Brass River crude oil. Poor corporate relations with indigenous communities, vandalism of oil infrastructure, severe ecological damage, and personal security problems throughout the Niger Delta oil-producing region continue to plague Nigeria's oil sector. Efforts are underway to reverse these troubles. In the absence of government programs, the major multinational oil companies have launched their own community development programs. A new entity, the Niger Delta Development Commission (NDDC), has been created to help catalyze economic and social development in the region. Although it has yet to launch its programs, hopes are high that the NDDC can reverse the impoverishment of local communities. The U.S. remains Nigeria's largest customer for crude oil, accounting for 40% of the country's total oil exports; Nigeria provides about 10% of overall U.S. oil imports and ranks as the fifth-largest source for U.S. imported oil.

The United Kingdom is Nigeria's largest trading partner followed by the United States. Although the trade balance overwhelmingly favors Nigeria, thanks to oil exports, a large portion of U.S. exports to Nigeria is believed to enter the country outside of the Nigerian government's official statistics, due to importers seeking to avoid Nigeria's excessive tariffs. To counter smuggling and under-invoicing by importers, in May 2001, the Nigerian government instituted a full inspection programme for all imports, and enforcement has been sustained. On the whole, Nigerian high tariffs and non-tariff barriers are gradually being reduced, but much progress remains to be made. The government also has been encouraging the expansion of foreign investment, although the country's investment climate remains daunting to all but the most determined. The stock of U.S. investment is nearly $7 billion, mostly in the energy sector. Exxon Mobil and Chevron are the two largest U.S. corporations in offshore oil and gas production. Significant exports of liquefied natural gas started in late 1999 and are slated to expand as Nigeria seeks to eliminate gas flaring by 2008.

Oil dependency, and the allure it generated of great wealth through government contracts, spawned other economic distortions. The country's high propensity to import means roughly 80% of government expenditures is recycled into foreign exchange. Cheap consumer imports, resulting from a chronically overvalued Naira, coupled with excessively high domestic production costs due in part to erratic electricity and fuel supply, pushed down utilization of industrial capacity to less than 30%. Many more Nigerian factories would have closed except for relatively low labor costs (10%–15%). Domestic manufacturers, especially pharmaceuticals and textiles, have lost their ability to compete in traditional regional markets. However, there are signs that some manufacturers have begun to improve competitiveness.

In April 2006, Nigeria became the first African country to fully pay off its debt owed to the Paris Club. However, this was structured as a debt writeoff of approximately $18 billion and a cash payment of approximately $12 billion. In the light of highly expansionary public sector fiscal policies in 2001, the government sought ways to head off higher inflation, leading to the implementation of stronger monetary policies by the Central Bank of Nigeria (CBN) and underspending of budgeted amounts. As a result of the CBN's efforts, the official exchange rate for the Naira has stabilized at about 112 Naira to the dollar. The combination of CBN's efforts to prop up the value of the Naira and excess liquidity resulting from government spending led the currency to be discounted by around 20% on the parallel (non-official) market. A key condition of the Stand-by Arrangement has been closure of the gap between the official and parallel market exchange rates. The Inter Bank Foreign Exchange Market (IFEM) is closely tied to the official rate. Under IFEM, banks, oil companies, and the CBN can buy or sell their foreign exchange at government influenced rates. Much of the informal economy, however, can only access foreign exchange through the parallel market. Companies can hold domiciliary accounts in private banks, and account holders have unfettered use of the funds.

Expanded government spending also has led to upward pressure on consumer prices. Inflation which had almost disappeared in April 2000 reached 14.5% by the end of the year and 18.7% in August 2001. In 2000, high oil prices resulted in government revenue of over $16 billion, about double the 1999 level. State and local governments demanded access to this "windfall" revenue, creating a tug-of-war between the federal government, which sought to control spending, and state governments desiring augmented budgets, preventing the government from making provision for periods of lower oil prices.

(c) Services

Nigeria ranks 63rd worldwide and fifth in Africa in services' output. Low power and telecom density has crippled the growth of this sector. Since undergoing severe distress in the mid-1990s, Nigeria's banking sector has witnessed significant growth over the last few years as new banks enter the financial market. Harsh monetary policies implemented by the Central Bank of Nigeria to absorb excess Naira liquidity in the economy has made life more difficult for banks, some of whom engage in currency arbitrage (round-tripping) activities that generally fall outside legal banking mechanisms. Private sector-led economic growth remains stymied by the high cost of doing business in Nigeria, including the need to duplicate essential infrastructure, the threat of crime and associated need for security counter measures, the lack of effective due process, and nontransparent economic decisionmaking, especially in government contracting. As of 2007, 29% of Nigerians in urban areas did not own bank accounts. While corrupt practices are endemic, they are generally less flagrant than during military rule, and there are signs of improvement. Meanwhile, since 1999 the Nigerian Stock Exchange has enjoyed strong performance, although equity as a means to foster corporate growth remains underutilized by Nigeria's private sector.

(d) Transport

Nigeria's publicly owned transportation infrastructure is a major constraint to economic development. Principal ports are at Lagos (Apapa and Tin Can Island), Port Harcourt, and Calabar. Docking fees for freighters are among the highest in the world. Of the 80,500 kilometers (50,000 mi.) of roads, more than 15,000 kilometers (10,000 mi.) are officially paved, but many remain in poor shape. Extensive road repairs and new construction activities are gradually being implemented as state governments, in particular, spend their portions of enhanced government revenue allocations. The government implementation of 100% destination inspection of all goods entering Nigeria has resulted in long delays in clearing goods for importers and created new sources of corruption, since the ports lack adequate facilities to carry out the inspection. Four of Nigeria's airports–Lagos, Kano, Port Harcourt and Abuja–currently receive international flights. Government-owned Nigeria Airways ceased operations in December 2002. Virgin Nigeria Airways started operations in 2005 as a replacement and serves domestic and international routes. There are several domestic private Nigerian carriers, and air service among Nigeria's cities is generally dependable. The maintenance culture of Nigeria's domestic airlines is not up to internationally accepted standards.

(e) Labour Forcer

In 2005, Nigeria had a labour force of 57.2 million. In 2003, the unemployment rate was 10.8 per cent overall; urban unemployment of 12.3 per cent exceeded rural unemployment of 7.4 per cent. According to the latest available information from 1999, labor force employment by sector was as follows: 70 per cent in agriculture, 20 per cent in services, and 10 per cent in industry. Labor unions, which have undergone periods of militancy and quiescence, reemerged as a force in 1998 when they regained independence from the government. Since 1999, the Nigerian Labour Congress (NLC), a union umbrella organization, has called six general strikes to protest domestic fuel price increases. However, in March 2005 the government introduced legislation ending the NLC's monopoly over union organizing. In December 2005, the NLC was lobbying for an increase in the minimum wage for federal workers. The existing minimum wage, which was introduced six years earlier but has not been adjusted since, has been whittled away by inflation to only US$42.80 per month. According to the International Organization for Migration, the number of immigrants residing in Nigeria has more than doubled in recent decades – from 477,135 in 1991 to 971,450 in 2005. The majority of immigrants in Nigeria (74 per cent) are from neighbouring Economic Community of West African States (ECOWAS), and that this number has increased considerably over the last decade, from 63 per cent in 2001 to 97 per cent in 2005. In spite of Nigeria's importance as a destination for migrants in the region, more people are emigrating from, than immigrating to, Nigeria with the negative net migration rate (per 1,000 people) steadily increasing in recent years, from -0.2 in 2000 to -0.3 in 2005, and this trend is expected to continue. According to recent estimates, the net migration rate could reach -0.4 in 2010.

(f) Gradual Reform

The Obasanjo government supports "private-sector" led, "market oriented" economic growth and has begun extensive economic reform efforts. Although the government's anti-corruption campaign has so far been disappointing, progress in injecting transparency and accountability into economic decisionmaking is notable. The dual exchange rate mechanism formally abolished in the 1999 budget remains in place in actuality. During 2000 the government's privatization program showed signs of life and real promise with successful turnover to the private sector of state-owned banks, fuel distribution companies, and cement plants. However, the privatization process has slowed somewhat as the government confronts key parastatals such as the state telephone company NITEL and Nigerian Airways. The successful auction of GSM telecommunications licenses in January 2001 has encouraged investment in this vital sector. Although the government has been stymied so far in its desire to deregulate downstream petroleum prices, state refineries, almost paralyzed in 2000, are producing at much higher capacities. By August 2001, gasoline lines disappeared throughout much of the country. The government still intends to pursue deregulation despite significant internal opposition, particularly from the Nigeria Labour Congress. To meet market demand the government incurs large losses importing gasoline to sell at subsidized prices.

4

Investment in Nigeria

4.1. Investment in Nigeria

Although Nigeria must grapple with its decaying infrastructure and a poor regulatory environment, the country possesses many positive attributes for carefully targeted investment and will expand as both a regional and international market player. Profitable niche markets outside the energy sector, like specialized telecommunication providers, have developed under the government's reform program. There is a growing Nigerian consensus that foreign investment is essential to realizing Nigeria's vast but squandered potential. European investments are increasing, especially since Belgian consultancy companies such as Genco are exploring the Nigerian market. Companies interested in long-term investment and joint ventures, especially those that use locally available raw materials, will find opportunities in the large national market. However, to improve prospects for success, potential investors must educate themselves extensively on local conditions and business practices, establish a local presence, and choose their partners carefully. The Nigerian Government is keenly aware that sustaining democratic principles, enhancing security for life and property, and rebuilding and maintaining infrastructure are necessary for the country to attract foreign investment. The stock market capitalisation of listed companies in Nigeria was valued at $97.75 billion on February 15, 2008 by the Nigerian Stock Exchange.

Nigeria's foreign economic relations revolve around its role in supplying the world economy with oil and natural gas, even as the country seeks to diversify its exports, harmonize tariffs in line with a potential customs union sought by the Economic Community of West African States (ECOWAS), and encourage inflows of foreign portfolio and direct investment. In October 2005, Nigeria implemented the ECOWAS common external tariff, which reduced the number of tariff bands.

Prior to this revision, tariffs constituted Nigeria's second largest source of revenue after oil exports. In 2005 Nigeria achieved a major breakthrough when it reached an agreement with the Paris Club to eliminate its bilateral debt through a combination of write-downs and buybacks. Nigeria joined the Organization of the Petroleum Exporting Countries in July 1971 and the World Trade Organization in January 1995.

4.1.1. Data

GDP: purchasing power parity – $459.4 billion (2009 est.)

GDP – real growth rate: 7% (July 2006 est.)

GDP – per capita: purchasing power parity – $3460 (2009 est.)

GDP – composition by sector:

- ✰ *Agriculture:* 26.8%
- ✰ *Industry:* 48.8%
- ✰ *Services:* 24.4% (2005 est.)

Population below poverty line: 78.98%(2000 est.)

Household income or consumption by percentage share:r

- ✰ *Lowest 10%:* 2.6%
- ✰ *Highest 10%:* 35.8% (1996–97)

Inflation rate (consumer prices): 7% (2006 est.)

Labor force: 57.21 million

Labor force – by occupation: agriculture 70%, industry 10%, services 20% (1999 est.)

Unemployment rate: 2.9% NA (2005 est.)

Budget:

- ✰ *Revenues:* $17 billion
- ✰ *Expenditures:* $13.54 billion including capital expenditures of $NA (2005 est.)

Industries: crude oil, coal, tin, columbite, palm oil, peanuts, cotton, rubber, wood, hides and skins, textiles, cement and other construction materials, food products, footwear, chemicals, fertilizer, printing, ceramics, steel, small commercial ship construction and repair.

Industrial production growth rate: 2.4% (2005 est.)

Electricity – production: 15.59 billion kWh (2003)

Electricity – production by source:

- ✰ *Fossil fuel:* 61.69%
- ✰ *Hydro:* 38.31%
- ✰ *Nuclear:* 0%
- ✰ *Other:* <.1% (1998)

Electricity - consumption: 14.46 billion kWh (2003)

Electricity - exports: 40 million kWh (2003)

Electricity - imports: 0 kWh (1998)

Oil - production: 2.35 million barrels per day (374×10^3 m^3/d) (July 2006 est.)

Oil - consumption: 310,000 bbl/d (49,000 m^3/d) (2003 est.)

Agriculture – products: cocoa, peanuts, palm oil, maize, rice, sorghum, millet, cassava (tapioca), yams, rubber; cattle, sheep, goats, pigs; timber; fish.

Exports: $72.16 billion f.o.b. (2005 est.)

Exports – commodities: petroleum and petroleum products 95%, cocoa, rubber

Exports – partners: United States 47.4%, Brazil 10.7%, Spain 7.1%(2004)

Imports: $45.95 billion f.o.b. (2005 est.)

Imports – commodities: machinery, chemicals, transport equipment, manufactured goods, food and live animals

Imports – partners: the People's Republic of China 9.4%, United States 8.4%, United Kingdom 7.8%, Netherlands 5.9%, France 5.4%, Germany 4.8%, Italy 4% (2004)

Debt – external: $3.3 billion with London Club(2006 est.)

Economic aid – recipient: IMF $250 million (1998)

Currency: 1 Naira (NGN) = 100 kobo

Exchange rates: Naira (NGN) per US$1 – 149.5 (2009), 120 (2006), 128 (2005), 132.89 (2004), 129.22 (2003), 120.58 (2002), 111.23 (2001)

External Reserves: $59 billion (2008)

Fiscal year: calendar year 2009

Economy of Nigeria

Currency: Nigerian naira (N) (NGN)

Trade organisations: OPEC

Statistics

GDP: $377.6 billion (2010 est.) (PPP; 31st)

GDP growth: 8.2% (2010 est.)

GDP per capita: $2,500 (2010 est.)

GDP by sector: agriculture: 41.8%; industry: 29.6%; services: 28.6% (2009 est.)

Inflation (CPI): 12.8% (2010 est.)

Population below poverty line: 45% (2010 est.)

Gini index: 43.7 (2003)

Labour force: 47.33 million (2009 est.)

Labour forceby occupation: agriculture: 70%; industry: 10%; services: 20% (1999 est.)

Unemployment: 4.9% (2007 est.)

Main industries: crude oil, coal, tin, columbite; palm oil, peanuts, cotton, rubber, wood; hides and skins, textiles, cement and other construction materials, food products, footwear, chemicals, fertilizer, printing, ceramics, steel, small commercial ship construction and repair, entertainment, machinery, car assembly.

Ease of Doing Business Rank: 133rd[1]

External

Exports: $45.43 billion (2009 est.)

Export goods: petroleum and petroleum products 95%, cocoa, rubber, machinery, processed foods, entertainment

Main export partners: United States 42%, Brazil 9.5%, India 9%, Spain 7.3%, France 5.1% (2008)

Imports: $42.1 billion (2009 est.)

Import goods: machinery, chemicals, transport equipment, manufactured goods, food and live animals

Main import partners: China 16.1%, Netherlands 11.3%, United States 9.8%, United Kingdom 6.2%, South Korea 6.1%, France 5.1%, Germany 4.4% (2008)

FDI stock: $71.59 billion (2009 est.)

Gross external debt: $9.689 billion (2009 est.)

Public Finances

Public debt: 17.8% of GDP (2009 est.)

Revenues: $10.49 billion

Expenses: $18.08 billion (2009 est.)

Credit rating: Standard & Poor's:

B+ (Domestic)

B+ (Foreign)

B+ (T&C Assessment)

Outlook: Stable Fitch:

BB-Outlook: Negative

Foreign reserves: $33 billion (2010; down from $46 billion in 2009 est.)

Main data source: CIA World Fact Book

All values, unless otherwise stated, are in US dollars

Source: UGANDA ECONOMY 2011.

The Ugandan economy recorded weaker growth of 5.1% in 2010 because of receding aggregate demand, mainly in private consumption, and weak external demand for traditional exports, in particular coffee. In spite of the declines, regional demand for Uganda's exports remained high. Export earnings fell from 2.9 billion US dollars (USD) in the financial year 2008/09 to USD 2.8 billion in 2009/10. Although lower than 2008/09 levels (USD 883 million), remittance receipts in 2009/10 (USD 820 million) surpassed traditional foreign exchange earners coffee and tourism. Earnings from coffee and tourism in 2009/10 were USD 262 million. The **economy of South Africa** is the largest in Africa, accounts for 24% of its Gross Domestic Product in terms of PPP, and is ranked as an upper-middle income economy by the World Bank, which makes the country one of only four countries in Africa represented in this category (the others being Botswana, Gabon and Mauritius). About a quarter of the population is unemployed and about the same proportion lives on less than US $1.25 a day.

Advanced development is significantly localised around four areas: *i.e.* Cape Town in the Western Cape, Port Elizabeth in the Eastern Cape, Durban in KwaZulu-Natal on the east coast, and Pretoria–Johannesburg in Gauteng Province. The latter province, although the smallest in land area of 17,010 square metres, accounts for 34% of the country's GDP and would it have been an independent territory, its economy would be larger than that of Morocco, Angola or Kenya. Beyond these four economic centres, development is marginal and poverty is still prevalent. However, key marginal areas have experienced rapid growth since the advent of democracy in 1994. Such areas include the Garden Route region from Mossel Bay to Plettenberg Bay; Rustenburg; Nelspruit or Mbombela; Bloemfontein; the Cape West Coast; and the KwaZulu-Natal North Coast.

The South African rand is the most actively traded emerging market currency in the world. It has joined an elite club of fifteen currencies, the continuous linked settlement (CLS), where forex transactions are settled immediately, lowering the risks of transacting across time zones. The rand was the best-performing currency against the United States dollar (USD) between 2002 and 2005, according to the Bloomberg Currency Scorecard.

4.2. Economic Systems Comparism

Focus Question: How does the U.S. economy compare to its competitors? (Case study: China)

Objectives

Students will be able to:

- ☆ Compare the U.S. to the Chinese economy.
- ☆ Discuss the positive and negative changes resulting from China's move to a more capitalist economy.
- ☆ Examine the life styles of people at different economic levels in China.

http://schools.nyc.gov/ve/eco/lesson20074.pdf

Note to Teachers

Selected section from The World is Flat by Thomas L. Friedman (Farrar, Straus and Giroux, 2005) needs to be read prior to the lesson.

Materials

- ☆ Textbook chapter on: Comparing Economic Systems
- ☆ *The World is Flat* by Thomas L. Friedman, "Chapter 2: The Ten Forces that Flattened the World, Flattener #6: Offshoring"
- ☆ Distribute prior to the lesson: Worksheet 4A, Worksheet 4C

Teaching Strategies

I. China vs. U.S. in 2000 and 2004

Distribute Worksheet 4A, "China vs. U.S. in 2000 and 2004 (Standard of Living)." Have students complete the exercise for homework prior to this lesson.

China vs. U.S. in 2000 and 2004* (Standard of Living)

	China		U.S.	
	2000	2004	2000	2004
Population	1.3 billion	1.3 billion	268.2 million	281.6 million
Life expectancy	70.3 years	71.4 years	77.1 years	77.4 years
GDP (total dollar value of goods and services produced in one year)	1.1 trillion	1.9 trillion	9.8 trillion	11.7 trillion
Inflation	2%	6.9%	2.2%	2.6%
Fixed lines and cell phones (per 1,000 people)	182	499.4	1,070	1,222.7
Internet users (per 1,000 people)	17.8	72.5	439	630.0
Merchandise trade** (% of GDP)	35.5	59.8	20.7	20.0

* Data from www.worldbank.org

** Merchandise trade as a share of GDP is the sum of merchandise exports and imports divided by the value of GDP.

Exercise: Your boss is going to China and wants you to write a memo in which you compare the U.S. and Chinese Standards of living. Write the memo in the space below:

To: The Boss

From: Me

Subject: Chinese vs. U.S. Standards of Living

On the day of the lesson, as part of the whole-class discussion, have students explain their answers to the following:

- ☆ What did you write in your memo?
- ☆ In your own words, how would you compare the standard of living in China with that of the U.S.?
- ☆ In what ways is the standard of living in China and the U.S. most similar? Most different?
- ☆ What would be the most useful pieces of information to a businessperson about China that appears on this chart?
- ☆ To what extent would you consider China a serious economic competitor to the U.S. today?

II. China's Economy: Four Views

Divide the class into groups of four. Distribute Worksheet 4B, "China's Economy: Four Views."

Have students follow the directions on the worksheet, and then have them explain the answers to the following: that China's leaders have vowed to cut millions of jobs, people like Mr. Wang continue to draw substantial salaries from the state. Mr. Wang works for the China Family Planning Association, a state-run organization that helps coordinate China's one-child policy with more than a million branches or lower-level organization across the country. He lives with his wife and son in a two-bedroom apartment for which he pays 230 yuan a month.

Mr. Wang's government salary is just 1,200 yuan a month, but he receives various benefits from hisemployer, like free lunches, that effectively increase his monthly income to about 2,000 yuan a month. His biggest expenses are his son's kindergarten, for which he pays 4,000 yuan a year, and yearly tuition of 10,000 for a two-year course he is taking towards a master's degree in social welfare.

Mr. Wang and his wife own a tiny car. Their most expensive possessions are a Motorola mobile telephone, desktop computer assembled from parts by a friend, and an aging color TV set. Thanks to Mr. Wang's wife's salary, the couple save about 3,000 yuan a month, and they have put about 60,000 yuan in the bank so far. China's economy remains largely state-owned, but its private sector is growing fast. There are now at least 10 million private business owners, not including millions more independent merchants, traders, and street hawkers. Xue Bing, 39, is typical of the new capitalist breed amassing wealth in the transitional economy. Mr. Xue was a manager of a state-owned construction company until two years ago, when it began to privatize many of its businesses. He invested 80,000 yuan, his life savings, to buy 8 per cent of one of the businesses, a company that manufactures paint. He owns a quarter stake in the company's Beijing subsidiary. Like most private businessmen, Mr. Xue spends much of his time cultivating relationships with officials to win his company business. He spends as much as 20,000 yuan a month on entertaining and gifts, though his company reimburses him for at least half of that. One of his biggest expenses is his daughter, for whom he spends about 20,000 yuan a year on school fees and an annuity. Mr. Xue draws a salary of 4,800 yuan a month and his wife earns 1,600 yuan a month as an office worker. But his annual income fluctuates because he can earn as much as 300,000 yuan for a single project. He has also made money on the stock market and today his savings total millions of yuan. His dream is to take control of the company and then take it public in about five years.

4.3. The Factory Worker

China has roughly 60 million factory workers, most of whom are still employed by the government. Many of these workers are losing their jobs, as a large number of the state-run factories close down. Tang Nianshan and his wife Tang Airong have spent most of their lives working in government-owned factories. Ms. Tang, 43, has been working for the same factory since she was 18. It makes enabled washbasins. She earns just under 1,000 yuan a month. Mr. Tang, 47, quit his factory job when the business began failing in the late 1980s as the government began withdrawing funding from the state-owned enterprises. Millions of people have lost their jobs in that transition, but Mr. Tang was lucky and found work in a Chinese-Thai joint venture that manufactures automobile parts. He earns 2,600 yuan a month on a welding line.

After work he goes home to a very small two-bedroom apartment that he and his family share with his parents. He and his wife sleep in one bedroom. Their 15-year-old son sleeps in the other with Mr. Tang's parents. His biggest possessions are his bicycle and a mobile telephone. The family has three TVs, one in each room. The family saves little money, but Shanghai's rapid development has filled Mr. Tang with hope, particularly for his son, whom he wants to go to college someday. "Our generation has seen many, many things," he says. "Today, life is getting better." *The New York Times*, 6/15/02, p. A6.

- ☆ What did you learn about the lives of the four Chinese people you read about?
- ☆ What did you learn about the Chinese economy?
- ☆ How do the lives of the Chinese you read about compare to the lives of Americans with similar positions?
- ☆ If your firm were interested in doing business in China, what kinds of information found in these four accounts would be most useful?

4.3.1. Offshoring

Discuss the following from Worksheet 4C, which was completed for homework:

- ☆ Define "offshoring."
- ☆ Why is China suited for offshoring?
- ☆ What effect does China's membership in WTO have on:
- ☆ American businesses?
- ☆ Other developing countries, such as Mexico and Vietnam?
- ☆ How does the author describe China's long-term strategy?
- ☆ What could disrupt China's successes?
- ☆ A study by Morgan Stanley published in Fortune magazine (10/4/04) found that this China flattener (offshoring) helped consumers. How?

- ☆ According to the reading, how might offshoring help the U.S. economy?
- ☆ Divide the class into four groups. Assign each group one of the following topics to discuss and present to the class:
- ☆ What does the author mean by "Either you get flat or you'll be flattened by China."
- ☆ According to the author, what effect does offshoring have, and will it continue to have, on the U.S. economy?
- ☆ You are running for office and your campaign is in favor of offshoring. Explain why.
- ☆ You are running for office and your campaign is against offshoring. Explain why.

4.3.2. Summary/Assessment

Ask students: "If you were a consultant to the Chinese government, what steps would you suggest the Chinese take to challenge the U.S. as the number one economic power in the world? Explain."

4.3.3. Homework

- ☆ Read materials in preparation for Lesson 5
- ☆ Assign questions for homework

China's Economy: Four Views

Assume that you work for the marketing department of your VE firm. Have one person from your group report on a different one of the four Chinese people below. In each report, you should discuss

- ☆ The kind of work the person does.
- ☆ How well the person lives.
- ☆ Whether or not you think the person would be a good target for the merchandise your firm sells.

4.3.4. The Farmer

Li Yongrong, 53, is a farmer. He says he earns about 5,000 yuan ($1 = 8.28 yuans) a year from the rice, wheat, canola, and pigs he has on four acres and leases from the state. But taxes and fees now eat up about 4,000 yuan of that income. School fees cost him another 2,000 yuan.

To make ends meet, he transports pigs and produce for fellow farmers in a Flying Tiger pickup truck thathe bought four years ago with 10,000 yuan borrowed from relatives and 3,000 yuan in savings. The trucking earns him enough to pay his electricity and telephone bills but there is barely enough leftover for the red envelopes of cash he is obliged to give as gifts each year during the Lunar New Yearholiday and at weddings and funerals. He saves almost nothing and remains deep in dept to relatives. Besides the truck and the telephone his grandest possession is a 21-inch color TV that he bought in 1996. Mr. Li and his wife, age 48, have no pension. Mr. Li wants his children to go to college. However, he does not know how he will pay for this even if they manage to pass the exams that allow just 13 per cent of China's young people to attend the state-run universities

Ghana Economic System

Economy of Ghana

Currency: Cedi (GHS)

Fiscal year: Calendar year

Statistics

GDP: $71.216 billion [1]

GDP growth: 14.337 % (2011 est.) [2]

GDP per capita: $2931 (2011 est.) [1]

GDP by sector: agriculture: 37.3%; industry: 25.3%; services: 37.5% (2006 est.)

Inflation (CPI): 8.1% (2011 est.)

Populationbelow poverty line: 28.5% (2007 est.)

Gini index: 39.4 (2005-06)

Labour force: 10.33 million (2009 est.)

Labour forceby occupation: agriculture: 56%; industry: 15%; services: 29% (2005 est.)

Unemployment: 11% (2000 est.)

Main industries: mining, lumbering, light manufacturing, aluminum smelting, food processing, cement, small commercial ship building

Ease of Doing Business Rank: 63rd[3]

External

Exports: $7.326 billion (2010 est.)

Export goods: gold, cocoa, timber, tuna, bauxite, aluminum, manganese ore, diamonds, horticulture

Main export partners: Netherlands 13.4%, Ukraine 11.7%, United Kingdom 8%, France 5.7%, United States 5.1% (2008)

Imports: $10.18 billion (2010 est.)

Import goods: capital equipment, petroleum, foodstuffs

Main import partners: China 16%, Nigeria 15%, India 5.6%, United States 5.6%, France 4.5%, United Kingdom 4.5% (2008)

Gross external debt: $5.84 billion (31 December 2009 est.)

Public Finances

Public debt: 67.5% of GDP (2009 est.)

Revenues: $4.547 billion (2009 est.)

Expenses: $6.124 billion (2009 est.)

Credit rating: Standard & Poor's:

B (Domestic)B (Foreign)

B+ (T&C Assessment)

Outlook: Negative

Fitch:

B+Outlook: Stable

Foreign reserves: $3.8 billion (31 December 2010 est.)

Main data source: CIA World Fact Book

All values, unless otherwise stated, are in US dollars

The **economy of Ghana**, West Africa, has a diverse and rich resource base, and as such, has one of the highest GDP per capita in Africa. Ghana remains somewhat dependent on international financial and technical assistance as well as the activities of the extensive Ghanaian diaspora. Gold, timber, cocoa, diamond, bauxite, and manganese exports are major sources of foreign exchange.[6] An oilfield which is reported to contain up to 3 billion barrels ($480\times10^{\Delta 6}$ m^3) of light oil was discovered in 2007.[7] Oil exploration is ongoing and, the amount of oil continues to increase.[8]

The domestic economy continues to revolve around subsistence agriculture, which accounts for 35% of GDP and employs 55% of the work force,[6] mainly small landholders. On the negative side, public sector wage increases and regional peacekeeping commitments have led to continued inflationary deficit financing, depreciation of the Cedi, and rising public discontent with Ghana's austerity measures. Furthermore, according to the World Bank, Ghana's per capita income has barely doubled over the past 45 years.[9] Even so, Ghana remains one of the more economically sound countries in all of Africa.

The country has, since July 2007, embarked on a currency re-denomination exercise, from Cedi (¢) to the new currency, the Ghana Cedi (GH¢). The transfer rate is 1 Ghana Cedi for every 10,000 Cedis. Ghana has embarked upon an aggressive media campaign to educate the public about what re-denomination entails. The new Ghana Cedi is now exchanging at a rate of $1 USD =Gh¢ 0.93.[citation needed]

Value Added Tax is a consumption tax administered in Ghana. The tax regime which started in 1998 had a single rate but since September 2007 entered into a multiple rate regime. In 1998, the rate of tax was 10% and amended in 2000 to 12.5%.

The country is mainly agricultural, with a majority of its workers engaged in farming. Ghana National Agricultural Export is the government arm that operates, maintains, overlooks the planting of cocoa, cashew etc. and other major crops for export. This agricultural arm of the government also harvests gold for export sales. Since its inception, it has drastically assisted the government in stabilizing the economy and boosting it as illegal sales of the major crops/nuts have been meaningfully and instantly curbed and also rendering employment to thousands of Ghanaians. Ghana's industrial base is relatively advanced compared to many other African countries. Import-substitution industries include textiles; steel (using scrap); tires; oil refining; flour milling; beverages; tobacco; simple consumer goods; and car, truck, and bus assembly. Ghana gets 97% of its energy from damming lake Volta, and exports much of this to neighboring countries tourism has become one of Ghana's largest foreign income earners (ranking third in 1997), and the Ghanaian Government has placed great emphasis upon further tourism support and development.

The financial services in Ghana has seen a lot of reforms in the past years. Ghana through the Banking (Amendment) Act 2007 has include the awarding of General Banking license to qualified Banks and this allows Offshore Banks to operate in the country. Barclays Bank (Ghana) limited has become the first Bank in Ghana to be awarded the General Banking license in the Country. It has therefore become possible for non-resident individuals and foreign companies to open offshore Bank Accounts in Ghana.

4.3.4. Economic History

Independence

At independence, Ghana had a substantial physical and social infrastructure and $481 million in foreign reserves. The Nkrumah government further developed the infrastructure and made important public investments in the industrial sector. With assistance from the United States, the World Bank, and the United Kingdom, construction of the Akosombo Dam was completed on the Volta River in 1966. Two U.S. companies built Valco, Africa's largest aluminium smelter, to use power generated at the dam. Aluminium exports from Valco were a major source of foreign exchange for Ghana. Many Nkrumah-era investments were monumental public works projects which were assets for the country, agricultural and industrial schemes. With cocoa prices falling and the country's foreign exchange reserves fast disappearing, the government resorted to supplier credits to finance many projects. By the mid-1960s, Ghana's reserves were gone, and the country could not meet repayment schedules. To rationalize, the National Liberation Council abandoned unprofitable projects, and some inefficient state-owned enterprises were sold to private investors. On three occasions, Ghana's creditors agreed to reschedule repayments due on Nkrumah-era supplier credits. Led by the United States, foreign donors provided import loans to enable the foreign exchange-strapped government to import essential commodities.

4.4. Acheampong Government

To restructure the economy, the NRC, under General Acheampong (1972–78), undertook an austerity program that emphasized self-reliance, particularly in food production. These plans were not realized, however, primarily because of post-1973 oil price increases and a drought in 1975-77 that particularly affected northern Ghana. The NRC, which had inherited foreign debts of almost $1 billion, abrogated existing rescheduling arrangements for some debts and rejected other repayments. After creditors objected to this unilateral action, a 1974 agreement rescheduled the medium-term debt on liberal terms. The NRC also imposed the Investment Policy Decree of 1975–effective on January 1977–that required 51% Ghanaian equity participation in most foreign firms, but the government took 40% in specified industries. Many shares were sold directly to the public.

4.4.1. Akuffo Government

Continued mismanagement of the economy, record inflation (more than 100% in 1977), and increasing corruption, notably at the highest political levels, led to growing dissatisfaction. The post-July 1978 military regime led by General Fred Akuffo attempted to deal with Ghana's economic problems by making small changes in the overvalued cedi and by restraining government spending and monetary growth. Under a one-year standby agreement with the International Monetary Fund (IMF) in January 1979, the government promised to undertake economic reforms, including a reduction of the budget deficit, in return for a $68 million IMF support program and $27 million in IMF Trust Fund loans. The agreement became inoperative, however, after the 4 June coup that brought Flight Lieutenant Jerry Rawlings and the AFRC to power for 3 months.

4.4.2. Limann Government

In September 1979, the civilian government of Hilla Limann inherited declining per capita income; stagnant industrial and agricultural production due to inadequate imported supplies; shortages of imported and locally produced goods; a sizable budget deficit (almost 40% of expenditures in 1979); high inflation, "moderating" to 54% in 1979; an increasingly overvalued cedi; flourishing smuggling and other black-market activities; unemployment and underemployment, particularly among urban youth; deterioration in the transport network; and continued foreign exchange constraints. Limann's PNP government announced yet another (2-year) reconstruction program, emphasizing increased food production and productivity, exports, and transport improvements. Import austerity was imposed and external payments arrears cut. However, declining cocoa production combined with falling cocoa prices, while oil prices soared. No effective measures were taken to reduce rampant corruption and black marketing. While it was waiting for realisations from its 2 year plan, the Limann government was interrupted by a Coup staged by Flt. Lt. Jeremiah John Rawlings.

4.4.3. Rawlings Government

When Rawlings again seized power at the end of 1981, cocoa output had fallen to half the 1970-71 level and its world price to one-third the 1975 level. By 1982, oil would constitute half of Ghana's imports, while overall trade contracted greatly. Internal transport had slowed to a crawl, and inflation remained high. During Rawlings' first year, the economy was stagnant. Industry ran at about 10% of capacity due to the chronic shortage of foreign exchange to cover the importation of required raw materials and replacement parts. Economic conditions deteriorated further in early 1983 when Nigeria expelled an estimated 1 million Ghanaians who had to be absorbed by Ghana. In April 1983, in coordination with the IMF, the PNDC launched an economic recovery program, perhaps the most stringent and consistent of its day in Africa, aimed at reopening infrastructural bottlenecks and reviving moribund productive sectors–agriculture, mining, and timber. The largely distorted exchange rate and prices were realigned to encourage production and exports. Increased fiscal and monetary discipline was imposed to curb inflation and to focus on priorities. Through November 1987, the cedi was devalued by more than 6,300%, and widespread direct price controls were substantially reduced.

5
Macro-economic Trend

5.1. Macro-economic Trend

This is a chart of trend of gross domestic product of Ghana at market prices estimated by the International Monetary Fund with figures in millions of Ghanaian Cedis.

Year	*Gross Domestic Product*	*US Dollar exchange*
1980	43,229	2.74 Cedis
1985	361,370	54.36 Cedis
1990	2,158,213	326.30 Cedis
1995	7,751,700	1,200.51 Cedis
2000	27,152,500	5,455.59 Cedis
2005	97,017,315	9,072.12 Cedis

The economy's response to these reforms was initially hampered by the absorption of one million returnees from Nigeria, the onset of the worst drought since independence, which brought on widespread bushfires and forced closure of the aluminium smelter and severe power cuts for industry and decline in foreign aid. In 1985, the country absorbed an additional 100,000 expellees from Nigeria. In 1987, cocoa prices began declining again; however, initial infrastructure repairs, improved weather, and producer incentives and support revived output in the early 1990s. During 1984-88 the economy experienced solid growth for the first time since 1978. Renewed exports, aid inflows, and a foreign exchange auction have eased hard currency constraints. Since an initial August 1983 IMF standby agreement, the economic recovery program has been supported by three IMF standbys and two other credits totaling $611 million, $1.1 billion from the World Bank, and hundreds of millions of dollars more from other donors. In November 1987, the IMF approved a $318-million, 3-year extended fund facility. The second phase (1987–90) of the recovery program concentrated on economic restructuring and revitalizing social services. The third phase, focused on financial transparency and macroeconomic stability is scheduled for March 1998.

Ghana intends to achieve its goals of accelerated economic growth, improved quality of life for all Ghanaians, and reduced poverty through macroeconomic stability, higher private investment, broad-based social and rural development, as well as direct poverty-alleviation efforts. These plans are fully supported by the international donor community and have been forcefully reiterated in the 1995 government report, Ghana: Vision 2020. Privatization of state-owned enterprises continues, with about two-thirds of 300 parastatal enterprises sold to private owners. Other reforms adopted under the government's structural adjustment program include the elimination of exchange rate controls and the lifting of virtually all restrictions on imports. The establishment of an interbank foreign exchange market has greatly expanded access to foreign exchange.

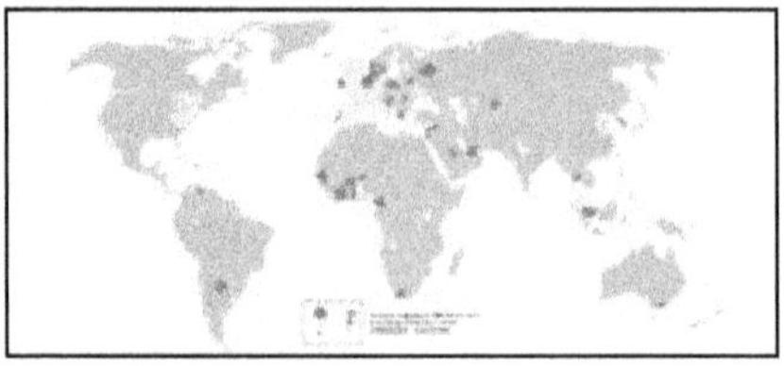

Ghanaian Exports in 2014

The medium-term macroeconomic forecast assumes political stability, successful economic stabilization, and the implementation of a policy agenda for private sector growth, and adequate public spending on social services and rural infrastructure. The ninth Consultative Group Meeting for Ghana ended 5 November 1997 after deliberations in Paris. Twenty-four countries and donor entities were represented at this meeting called by the World Bank on behalf of the Ghanaian Government. The World Bank announced that, of the targeted disbursement level of $1.6 billion sought from the donor community for 1998-99, they foresaw only a $150 million shortfall in commitments, and that this shortfall would be easily realized should Ghana rapidly enact its macroeconomic program. The government repealed a 17% value-added tax (VAT) shortly after its introduction in 1995, which resulted in widespread public protests. The government reverted to several previously imposed taxes, including a sales tax. The government has set in motion a program to reintroduce a VAT bill, with implementation in 1998 after an extensive public education campaign.

5.2. Comparative Economic System in Africa

Colonial rule in Africa was intended to be cheap, *viz.* for taxpayers in Europe. The British doctrine was that each colony should be fiscally self-supporting. Thus, any growth in government expenditure was supposed to be financed from higher revenues, as it was in Ghana in the 1920s when Governor Guggisberg was able to fund the creation of what became the country's best-known hospital and school, as well as a new harbour and more railways and roads, from customs proceeds that had been fuelled by the colony's increasing exports of cocoa beans. In practice the French were equally committed to covering costs. In French West Africa too there was a major programme of public works in the 1920s, although, as also in Ghana, within a few years expenditure had to be curtailed when export prices fell and the growth of revenue ended (Hopkins 1973, 190).

After retrenchment during the 1930s Depression, and especially during the Second World War, colonial administrations found themselves (for a variety of reasons) entering the post-war era with a new public commitment to be seen to promote actively the development of the economies over which they presided. "Developmental" language was partly redeemed by greater spending. In principle this came partly from the metropolitan taxpayer. However, in the French case Patrick Manning (1998, 123-5) has calculated that the government continued to receive more in tax from Africa than it spent in Africa. In British West Africa the new statutory export marketing boards accrued substantial surpluses by keeping a large margin between the price paid to producers and the price that the boards received for the crop on the world market. The surpluses were kept in London, in British government bonds, as forced savings from African farmers (Rimmer 1992, 41-2), which assisted the British metropolitan economy to recover from the post-war dollar shortage.

The particular identity of the colonial power made some difference to the lives of those subjected to European rule. Contrasts between the two largest empires in Africa are traditionally made with reference to greater British reliance on African chiefs as intermediaries ("indirect rule") and the French doctrine of assimilating a small minority of Africans into French culture and citizenship. On the whole it is arguable that, in economic terms, the similarities were much greater than the differences, except when the latter arose from the composition of their respective African empires. French rule, like British, relied on African intermediaries, including chiefs, even though France was much more insistent on abolishing African monarchies (as in Dahomey, in contrast to the British treatment of the structures and dynasties of the States of Buganda, Botswana, Lesotho and, after an abortive attempt at abolition, Ashanti). In West Africa the French made much greater use of forced labour, but that was primarily because the French territories were, from the start, relatively lacking in cash-earning and therefore wage-paying potential. That specific policy, *Corvée* and its use to benefit white planters rather than African farmers, made a difference to the colonial legacy in Ghana and Côte d'Ivoire. It meant that African cocoa farming took off much more quickly and dramatically in the former, so that Ghana was much wealthier at independence, when Côte d'Ivoire was in the process of catching up (and overtaking) after a late start (Hopkins 1973, 218-9), which it proceeded to do by the 1980s.

☆ This is also the case for Madagascar, which is beyond the scope of the discussion here.

☆ Gross national incomes in PPP terms averaged USD 918 in the former British colonies and USD 1,208.

The contention here that the differences between the legacies of British and French rule in Africa are primarily attributable to variations in the composition of the African empires concerned may need to be qualified in the light of valuable recent research by Thomas Bossuroy and Denis Cogneau (2009). They examined social mobility in five African countries and found that in the former British colonies in their sample, Ghana and Uganda, "the links between origin, migration, education and occupational achievement appear much looser" than in the former French colonies they examined, *i.e.* Côte d'Ivoire and Guinea3 (Bossuroy and Cogneau 2009, 2). In explanation, they emphasise the importance of greater investment in education in the British colonies than the French colonies in their sample. This is a novel and important line of inquiry. I suspect that the favourable conclusion about the former British colonies also partly reflects the fact that Ghana and Uganda, for reasons that are only partly and indirectly connected to their respective British legacies, were the two major growth successes of structural adjustment in Africa, therefore perhaps offering greater opportunities for educational, physical and occupational mobility from the mid-1980s, which was early enough to be partly reflected in the data and which coincided with economic stagnation and then civil war in Côte d'Ivoire. Finally, the contrast may also partly reflect the legacy of the era of *Corvée* and "settler" agriculture in Côte d'Ivoire, before the economy took off in the 1950s and 1960s. As of 1990, helped by the legacy of the Ivorian "miracle", in a sample of 26 former British and French colonies in tropical Africa (so excluding southern Africa) it was the former French colonies that had the higher per capita incomes in Purchasing Power Parity (PPP) terms, by over 30% (Bossuroy and Cogneau 2009, 45, citing World Bank data).

5.3. Colonial Rule and Africa's Specialisation in Primary Product Exports

The "extraversion" and "monoculture" of African economies is widely lamented and condemned as a victory of colonial interests over African interests. The risks entailed in extreme specialisation, however, need to be set against the long-run income gain to be expected from the exploitation of comparative advantage. But again, although the location of a colonial economy's comparative advantage could be identified, sooner or later the task of capitalising on it raised the question of what investments might profitably deepen that advantage and, above all, of how the costs and benefits would be distributed. Conflicts of ideology, and especially the balance of power between different interest groups, worked out variously across the range of African colonies. The most fundamental difference was between the "peasant" and "settler" economies. Let us consider the contrasting cases of export agriculture in the former, notably in West Africa, and mining in the latter, most obviously in South Africa.

We have noted that, by the eve of the European partition of the continent, Africa had already revealed an emerging comparative advantage in export agriculture. In West Africa in particular it was in the joint interests of the population, European merchants and the colonial administrations to further this. In Ghana British planters were initially allowed to enter to grow cocoa beans. But lacking the discriminatory support from the government that their counterparts enjoyed in Kenya and southern Africa, they failed in commercial competition with African producers (Austin 1996a), just as French planters were later to be eclipsed by African ones in Côte d'Ivoire following the abolition of *Corvée*. Colonial reliance on the efforts of African small capitalists and peasants in the growing and local marketing of export crops paid off in what became Ghana and Nigeria, with more than 20-fold increases in the real value of foreign trade between 1897 and 1960 (Austin 2008a, 612), benefiting British commercial interests as well as (via customs duties) the colonial treasury. The efforts of W. H. Lever, the soap manufacturer, to win government permission, along with the necessary coercive support, to establish huge oil palm plantations in Nigeria continued from 1906 to 1925, but they were always rebuffed in favour of continued African occupation of virtually all agricultural land. Ultimately this was because African producers literally delivered the goods (Hopkins 1973, 209-14) through land-extensive methods well adapted to the factor endowment. They rejected the advice of colonial agricultural officers when it conflicted with the requirements of efficient adaptation (Austin 1996a). The positive contribution of the administrations was to reinforce and permit the exploitation of these economies' comparative advantage in export agriculture. They did this partly by investment in transport infrastructure, investment to which African entrepreneurs also contributed (Austin 2007). Equally important, although the colonial administration never really established a system of land titling, in Ghana (for example) it upheld the indigenous customary right of farmers to ownership of trees they had planted, irrespective of the outcome of any later litigation about the ownership of the land the trees stood upon. Thus, African producers enjoyed sufficient security of tenure to feel safe in investing in tree crops on a scale sufficient to create, in the case of Ghana, what became for nearly 70 years the world's largest cocoa economy (Austin, 2005).

South Africa had gold and diamonds, but their profitable exploitation required that the cost of labour be reduced far below what the physical labour-land ratio implied. C. H. Feinstein's quantitative exercise indicates that without such coercive intervention in the labour market, most of South Africa's mines would have been unprofitable until the end of the gold standard era in 1932 (Feinstein 2005, 109-12). If South Africa eventually obtained a "free market" comparative advantage in mining, it was only after several decades of using extra-market means to repress black wages, notably through land appropriation and measures to stop Africans from working on European-owned land except as labourers rather than tenants.

Comparison of the economic legacies of European rule for poverty in "settler" and "peasant" economies is complicated by the many variations between individual colonies. However, some generalisations are possible. It is clear that the distribution of wealth and income was, and has remained, much more unequal in the "settler" economies than in the "peasant" ones. Preliminary findings by Sue Bowden, Blessing Chiripanhura and Paul Mosley (2008) support the proposition that possession of land put a floor under real wages in the "peasant" colonies, enabling labourers migrating into export-crop growing areas to share in the gains from exports that were otherwise divided between European firms, African and Asian middlemen and African farm-owners (see also Austin 2005). Bowden, Chiripanhura and Mosley find real wages beginning to rise from the 1920s and 1930s in the "peasant" colonies of Ghana and Uganda respectively and not falling back afterwards to the 1914 floor. In contrast, it was only in the 1970s that the real wages of black gold-miners in South Africa began a sustained rise above their early 20th century level (Lipton 1986, 410). In the sample taken by Bowden, Chiripanhura and Mosley it was only in the "pure settler" economies, South Africa and Zimbabwe (Southern Rhodesia), not in the "peasant" colonies of Ghana or Uganda nor even in the intermediate case of Kenya, that there were declines in rural African living standards over periods of longer than 15 years within the 20th century. This pattern in real wages, together with the long-term expansion of African export agriculture which underpinned the growth of real wages in the "peasant" colonies, was reflected in the earlier onset of falling infant mortality in Ghana and Uganda compared to Southern Rhodesia and South Africa.

It should be added that many African colonies were short of both known mineral deposits and the kinds of land suitable for profitable export agriculture. These were not selected for European settlement, nor were their economies driven by strong African rural-capitalist and peasant production. They had to rely on seasonal exports of male wage labourers, and on growing the less lucrative cash crops such as cotton, the timing of whose labour requirements conflicted with those of food crops, thereby creating risks to food security (Tosh 1980). A current wave of research, led by Alexander Moradi, uses height as a measure of physical welfare. The average height of African populations rose during the colonial period in Ghana and even in the "semi-settler" economy of Kenya (Moradi 2008; 2009). When this research is extended to poorer colonies such as southern Sudan, Tanganyika (mainland Tanzania) or those in the West African Sahel, it would be no surprise if welfare improvements there are found to have been smaller than in the better-endowed economies studied so far. It was particularly in (selected areas of) the less favourably-endowed economies that colonial governments sought to raise productivity through very large-scale, capital-intensive and authoritarian projects, notably the massive irrigation scheme of the Office du Niger in Mali and the mechanisation campaign of the East African Groundnut Project in Tanganyika. Both were spectacular failures in their own output and productivity terms, not least because they were inefficient in relation to the prevailing factor ratios and physical environments (Hogendorn and Scott 1981; Roberts 1996, 223-48; Van Beusekom 2002).

Poor as was the record of "settler" colonialism for the living standards of the indigenous population, it was in colonies where Europeans appropriated land on a large scale, for settlers or for companies, that the earlier and larger beginnings were made in modern manufacturing

5.4. Markets and African Entrepreneurship

African entrepreneurship has driven changes in the choice of products and in the means and organisation of production in various contexts before, during and since colonial rule.3 This has been particularly conspicuous in West Africa, whose

19th century pre-colonial economies tended to be regarded as more market-oriented than those of the other major regions of Sub-Saharan Africa (Austin, forthcoming). The colonial impact on African entrepreneurship and on the markets in which they operated again turned to a large extent on whether there were large-scale appropriations of land for the use of Europeans, be they individual settlers or corporations. This familiar division between "settler" and "plantation" colonies, on one hand, and "peasant" (and rural capitalist) colonies, on the other, was far from purely exogenous to African economic history. Where African producers were able to enter export markets early and on a wide scale, before European exporters really got going, their success was sufficient to tip the balance of the argument among colonial policy-makers in favour of those who thought it economically as well as politically wisest to leave agricultural production in African hands. British West Africa was the major example of this. In South Africa, Southern Rhodesia and Kenya African farmers responded quickly to opportunities to grow additional grain to supply internal markets. But the governments reacted by trying to drive Africans out of the produce market and into the labour market by reserving land for Europeans, while either prohibiting Africans from leasing it back or (as in inter-war Kenya) limiting the time that African "squatters" could work for themselves rather than for their European landlords (Palmer and Parsons 1977; Kanogo 1987). African production for the market proved resilient, however, and the governments eventually accepted this and shifted to imposing controls on agricultural marketing that favoured European producers rather than trying to displace African ones. In Kenya it was only in the mid-1950s, during the Mau Mau revolt, that the government lifted restrictions on African production of high-value cash crops (Mosley 1983). Thus, to the extent that African production for the market in the late 19th century was greater in what became the "peasant" agricultural-export economies than in what became the "settler" economies, that contrast was reinforced by government actions in the latter over the following decades. Not that the maintenance of African ownership of land necessarily entailed support for African capitalism. Admittedly, we have seen that the colonial State in Ghana protected the property of agricultural investors, in the sense of preserving the ownership of a farmer over trees or crops that he or she had planted, irrespective of the outcome of legal disputes about the ownership of the land on which they stood. But in "settler" and "peasant" colonies alike colonial governments were hesitant and usually hostile to the emergence of land markets in areas controlled by Africans. This policy eventually changed in Southern Rhodesia and Kenya, with selective promotion of land registration, in response to the *de facto* emergence of land sales and individual proprietorship (cultivable land having become increasingly scarce in the areas left to Africans) and with African land-owners being seen as a politically conservative force in the context of Mau Mau (Mosley 1983, 27-8; Kanogo 1987). In West Africa, without the settler pressure on African access to land, and given the expansion of cash crops that occurred early in the colonial period and again in the 1950s, neither the political case nor the economic case for compulsory land titling was as yet compelling (Austin, 2005).

African entrepreneurs, like European ones, needed to be able to hire labour. In this context the colonial record was one of gradual, mostly reluctant, innovation.

Sooner or (often) later, they legislated against slavery. But in West Africa, the region with evidently the largest slave population at the start of the 20th century, the replacement of the slave market by a wage labour market depended very much on the progress of African cash crop agriculture (Austin 2009). During the inter-war decades the continued use of forced labour by colonial administrations came under sustained pressure from the International Labour Office in Geneva. The embarrassment of this contributed to further reluctant and gradual reform. By the end of the Second World War, as Frederick Cooper (1996) has shown, British and French authorities had accepted that wage labour had become a regular occupation for Africans, rather than a seasonal sideline from farming. Indeed, Cooper went on to show that in London and Paris the long-run fiscal implications of having to give workers in Africa the same rights as workers in Europe contributed to the decisions to withdraw from tropical Africa. For African societies the end of slavery and the rise of wage labour was arguably a condition of continued large-scale participation in international trade. As early as 1907 the chocolate manufacturer Cadbury had moved its cocoa-buying to Ghana following bad publicity about "slave-grown" cocoa in the Portuguese colony of Sao Tome, where it had been buying before (Southall 1975, 39-49). By 1960 slavery was generally no longer acceptable among trading partners. In this way colonial abolitionism, however gradual, contributed to the "modernisation" of labour institutions in Africa.

Colonial rule facilitated the import of capital into this capital-scarce continent. But only in mining, and to some extent in "settler" and "plantation" agriculture, did this happen on a large scale. The survey by Herbert S. Frankel (1938) of external capital investment in white-ruled Africa remains the only comprehensive study for the colonial period. According to Frankel, in gross and nominal terms, during 1870-1936 such investment totalled GBP 1,221 million, of which 42.8% was in South Africa. This meant GBP 55.8 per head in South Africa, but only GBP 3.3 in the French colonies and GBP 4.8 in British West Africa. Public investment constituted 44.7% of the grand total and almost 46% of the non-South Africa total (Frankel 1938, 158-60, 169-70). Governments, and to some extent mining and plantation companies, invested in the transport infrastructure required for the development of, mainly, the export-import trade. In Nigeria and Ghana Africans also took a leading role in building motor roads and pioneering lorry services (Heap 1990). In institutional terms the colonial period saw the eventual abolition of human pawning, with its replacement by promissory notes and, in those areas of West Africa where it was possible, by loans on the security of cocoa farms. It also saw the introduction of modern banking, but the banks were much more willing to accept Africans' savings than to offer them loans, partly because of the colonial governments' non-introduction of compulsory land titling (Cowen and Shenton, 1991).

Like shipping and the export-import trade, banking in colonial West Africa had a strong tendency towards cartelisation (Olukoju 2001-02; Austin and Uche 2007). The initial imposition of colonial rule and boundaries itself disrupted intra-African networks of exchange, and the increasing presence of European merchants in the interior relegated many African traders further down the chain of intermediation between shippers and farmers (Goerg 1980; Nwabughuogu, 1982). Organised resistance to European cartels mostly emerged later and was largely confined to particular colonies, given the tradition of cocoa "hold-ups" with which African farmers and brokers confronted successive European merchant cartels in Ghana and the indigenous banking movement in Nigeria (Miles 1978; Hopkins 1966). Until the advent of independence it remained the case in the "peasant" colonies that the markets dominated by Europeans were cartelistic, whereas the markets populated by Africans were characterised by extreme competition (Hopkins 1978, 95). At least, much more than in the "settler" colonies, African entrepreneurs were able to operate in the export-import as well as in the domestic exchange sectors. Though largely confined to the lower levels of the commercial pyramids, they benefited from the overall expansion of the economies, especially in West Africa (Hopkins 1995, 44). The monopolistic arrangements were shaken a bit by decolonisation (Austin and Uche 2007), but the old African sector became the "informal" sector, the old European sector the "formal" one (Austin, 1993). At independence, new governments were faced with the familiar problems of this financial dualism, notably lack of cheap credit from the formal sector for informal enterprises.

Despite asymmetric competition, the more economically successful "peasant" colonies saw the continuation of a tradition of entrepreneurship and mostly (but not always) small-scale accumulation in agriculture, crafts and trade. The result, as John Iliffe (1983, 67) noted, was "a strong contrast between West Africa, with its long-established capitalistic sector and its entrepreneurs from artisanship and trade, and eastern and southern Africa, where entrepreneurs had emerged chiefly through Western education and modern sector employment". Early post-colonial policy did not always build on this, for example in the case of Ghana, with high taxation of export agriculture and the creation of State monopolies in certain sectors (Austin, 1996b).

6

Major Obstacle to Nigeria Economic Growth

6.1. Major Obstacle to Nigeria Economic Growth

In Nigeria, high transport prices are rightly seen as a major obstacle to economic growth of the region. This has been documented in several studies. One of them (Amjadi and Yeats, 1995) even concluded that, in Nigeria, transport costs are a higher trade barrier than import tariffs and trade restrictions. In this paper, it is analyzed, identified and quantified the factors behind Nigeria's high prices of road transport, with the purpose of helping policy-makers adopt appropriate actions to reduce transport prices to domestic and international trade. Before the presentation of the results, a distinction has to be made between logistics costs, transport prices, transport costs and vehicle operating costs in order to avoid the usual confusion. This distinction is useful because prices may or may not reflect transport cost:

- Transport prices are the rates charged by a transport company or a freight forwarder to the shipper or importer. Transport prices usually are the result of negotiated rates between the shipper and the transport service provider. Transport prices normally cover transport costs, the operator's overheads and profit margin.
- Transport costs are the costs the transport operator incurs when transporting a cargo. In addition to VOC, transport cost includes other indirect costs, such as license fee, roadblocks, etc.

- Vehicle operating costs (VOC) include the various direct costs the transport provider must pay to operate a given vehicle, notably labor, capital, fuel, tyres, maintenance and depreciation cost of a vehicle.

However, a number of studies that looked at the determinants of the high transport costs in Africa did not arrive at a wide consensus, but rather identified different factors as being the main contributors to the high prices. The factors included: low productivity of the trucking industry in Africa, notably due to infrastructure constraints (Pedersen, 2001), low levels of competition between service providers (Rizet and Hine, 1993), or weak infrastructure (Limao and Venables, 2001). This latter study also indicated the poor quality of infrastructure as accounting for most of Africa's poor trade performance. From a cross country regression, the study concluded that trade was highly sensitive to transport costs – a 10 per cent drop in transport costs increases trade by 25 per cent, and transport costs are sensitive to the quality of infrastructure, as measured by such variables as the density of the road and rail network. In the past, large investments in improving road infrastructure were predicated on the basis that transport prices would be reduced. The World Bank since the 1970s have actively supported the transport corridors in Africa and during many years such support focused almost exclusively on improving infrastructure. While the improvements were essential to facilitate road transport and resulted in lower costs for the trucks carrying cargo on the corridors, no clear impact on the transport prices was evident. Thus, the end-users of road transport services did not seem to fully benefit from the lower transport costs and better service quality that the improved infrastructure could possibly allow. Since many research attempts in the past did not arrive at a conclusive result, the current study has made a deliberate attempt to expand both the breadth and depth of the research and can claim to be original in the following areas:

- It is the first comprehensive and practical effort in the last fifteen years to measure and quantify the high transport prices in Nigeria.
- It is the first attempt of this kind in Nigeria and elsewhere in the world to disaggregate and clearly define the various input factors into three tiers of cost and prices, *i.e.* vehicle operating costs, transport costs (incurred to commercial transport providers) and transport prices (incurred to end users), and try to understand how each of them effect the final transport prices.
- It is also the first study that clearly recognizes the diversification in transportation market across Africa continent with an attempt to measure with the same yard stick the costs, prices and performances of the transportation industry across the four sub-regions of Africa. Contrary to what most literature has presented, transport costs are not excessively high in Africa but transport prices are, especially in Central Africa mainly due to official and unofficial market regulation and structure in trucking services.

6.1.1. Transport Costs and Rates

Transport systems face requirements to increase their capacity and to reduce the costs of movements. All users (*e.g.* individuals, enterprises, institutions, governments, etc.) have to negotiate or bid for the transfer of goods, people, information and capital because supplies, distribution systems, tariffs, salaries, locations, marketing techniques as well as fuel costs are changing constantly. There are also costs involved in gathering information, negotiating, and enforcing contracts and transactions, which are often referred as the cost of doing business. Trade involves transactions costs that all agents attempt to reduce since transaction costs account for a growing share of the resources consumed by the economy. Frequently, enterprises and individuals must take decisions about how to route passengers or freight through the transport system. This choice has been considerably expanded in the context of the production of lighter and high value consuming goods, such as electronics, and less bulky production techniques. It is not uncommon for transport costs to account for 10% of the total cost of a product. This share also roughly applies to personal mobility where households spend about 10% of their income for transportation, including the automobile which has a complex cost structure. Thus, the choice of a transportation mode to route people and freight within origins and destinations becomes important and depends on a number of factors such as the nature of the goods, the available infrastructures, origins and destinations, technology, and particularly their respective distances. Jointly, they define transportation costs.

Transport costs are a monetary measure of what the transport provider must pay to produce transportation services. They come as fixed (infrastructure) and variable (operating) costs, depending on a variety of conditions related to geography, infrastructure, administrative barriers, energy, and on how passengers and freight are carried. Three major components, related to transactions, shipments and the friction of distance, impact on transport costs. Transport costs have significant impacts on the structure of economic activities as well as on international trade. Empirical evidence underlines that raising transport costs by 10% reduces trade volumes by more than 20%. In a competitive environment where transportation is a service that can be bided on, transport costs are influenced by the respective rates of transport companies, the portion of the transport costs charged to users.

Rates are the price of transportation services paid by their users. They are the negotiated monetary cost of moving a passenger or a unit of freight between a specific origin and destination. Rates are often visible to the consumers since transport providers must provide this information to secure transactions. They may not necessarily express the real transport costs. The difference between costs and rates either results in a loss or a profit from the service provider. Considering the components of transport costs previously discussed, rate setting is a complex undertaking subject to constant change. For public transit, rates are often fixed and the result of a political decision where a share of the total costs is subsidized by the society. The goal is to provide an affordable mobility to the largest possible segment of the population even if this implies a recurring deficit (public transit systems rarely make any profit). It is thus common for public transit systems to have rates that are lower than costs. For freight transportation and many forms of passenger transportation (*e.g.* air transportation) rates are subject to a competitive pressure. This means that the rate will be adjusted according to the demand and the supply. They either reflect costs directly involved with shipping (cost-of-service) or are determined by the value of the commodity (value-of-service). Since many actors involved in freight transportation are private rates tend to vary, often significantly, but profitability is paramount.

6.1.2. Costs and Time Components

Transportation offers a spectrum of costs and level of services. Among the most significant conditions affecting transport costs and thus transport rates are:

- **Geography**. Its impacts mainly involve distance and accessibility. Distance is commonly the most basic condition affecting transport costs. The more it is difficult to trade space for a cost, the more the friction of distance is important. It can be expressed in terms of length, time, economic costs or the amount of energy used. It varies greatly according to the type of transportation mode involved and the efficiency of specific transport routes. Landlocked countries tend to have higher transport costs, often twice as much, as they do not have direct access to maritime transportation.
- **Type of product**. Many products require packaging, special handling, are bulky or perishable. Coal is obviously a commodity that is easier to transport than fruits or fresh flowers as it requires rudimentary storage facilities and can be transshipped using rudimentary equipment. Insurance costs are also to be considered and are commonly a function of the value to weight ratio and the risk associated with the movement. As such, different economic sectors incur different transport costs as they each have their own transport intensity. With containerization the type of product plays little in the transport cost since rates are set per container, but products still need to be loaded or unloaded from the container. For passengers, comfort and amenities must be provided, especially if long distance travel is involved.
- **Economies of scale**. Another condition affecting transport costs is related to economies of scale or the possibilities to apply them as the larger the quantities transported, the lower the unit cost. Bulk commodities such as energy (coal, oil), minerals and grains are highly suitable to obtain lower unit transport costs if they are transported in large quantities. For instance, moving a barrel of oil over 4,000 km would cost $1 on a 150,000 deadweight tons tanker ship and $3 on a 50,000 deadweight tons tanker ship. A similar trend also applies to container shipping with larger containerships involving lower unit costs.
- **Energy**. Transport activities are large consumers of energy, especially oil. About 60% of all the global oil consumption is attributed to transport activities. Transport typically account for about 25% of all the energy consumption of an economy. The costs of several energy intensive

transport modes, such as air transport, are particularly susceptible to fluctuations in energy prices.

- **Trade imbalances.** Imbalances between imports and exports have impacts on transport costs. This is especially the case for container transportation since trade imbalances imply the repositioning of empty containers that have to be taken into account in the total transport costs. Consequently, if a trade balance is strongly negative (more imports than exports), transport costs for imports tend to be higher than for exports. Significant transport rate imbalances have emerged along major trade routes. The same condition applies at the national and local levels where freight flows are often unidirectional, implying empty movements.
- **Infrastructures.** The efficiency and capacity of transport modes and terminals has a direct impact on transport costs. Poor infrastructures imply higher transport costs, delays and negative economic consequences. More developed transport systems tend to have lower transport costs since they are more reliable and can handle more movements.
- **Mode.** Different modes are characterized by different transport costs, since each has its own capacity limitations and operational conditions. When two or more modes are directly competing for the same market, the outcome often results in lower transport costs. Containerized transportation permitted a significant reduction in freight transport rates around the world.
- **Competition and regulation.** Concerns the complex competitive and regulatory environment in which transportation takes place. Transport services taking place over highly competitive segments tend to be of lower cost than on segments with limited competition (oligopoly or monopoly). International competition has favored concentration in many segments of the transport industry, namely maritime and air modes. Regulations, such as tariffs, cabotage laws, labour, security and safety impose additional transport costs, particularly in developing countries.
- **Surcharges.** Refer to an array of fees, often set in an arbitrary fashion, to reflect temporary conditions that may impact on costs assumed by the transporter. The most common are fuel surcharges, security fees, geopolitical risk premiums and additional baggage fees.

The transport time component is also an important consideration as it is associated with the service factor of transportation. They include the transport time, the order time, the timing, the punctuality and the frequency. For instance, a maritime shipper may offer a container transport service between a number of North American and Pacific Asian ports. It may take 12 days to service two ports across the Pacific (transport time) and a port call is done every two days (frequency). In order to secure a slot on a ship, a freight forwarder must call at least five days in advance (order time). For a specific port terminal, a ship arrives at 8AM and leaves at 5PM (timing) with the average delay being two hours (punctuality).

Mobility tends to be influenced by transport costs. Empirical evidence for passenger vehicle use underlines the relationship between annual vehicle mileage and fuel costs, implying the higher fuel costs are, the lower the mileage. At the international level, doubling of transport costs can reduce trade flows by more than 80%. The more affordable mobility is, the more frequent the movements and the more likely they will take place over longer distances. A wide variety of transport costs can be considered.

Freight on board (FOB). Is a transport rate where the price of a good is the combination of the factory costs and the shipping costs from the factory to the consumer. In the case of FOB, the consumer pays for the freight transport costs. Consequently, the price of a commodity will vary according to transportation costs and distance.

Costs-Insurance-Freight (CIF). Is a transport rate that considers the price of the good, insurance costs and transport costs. It implies a uniform delivered price for all customers everywhere, with no spatially variable shipping price. The average shipping price is built into the price of a good. The CIF cost structure can be expanded to include several rate zones, such as one for local, another for the nation and another for exports.

Terminal costs. Costs that are related to the loading, transshipment and unloading. Two major terminal costs can be considered; loading and unloading at the origin and destination, which are unavoidable, and intermediate (transshipment) costs that can be avoided.

Linehaul costs. Costs that are a function of the distance over which a unit of freight or passenger is carried. Weight is also a cost function when freight is involved. They include labor and fuel and commonly exclude transshipment costs.

Capital costs. Costs applying to the physical assets of transportation mainly infrastructures, terminals and vehicles. They include the purchase or major enhancement of fixed assets, which can often be a one-time event. Since physical assets tend to depreciate over time, capital investments are required on a regular basis for maintenance.

Transport providers make a variety of decisions based on their cost structure, a function of all the above types of transport costs. The role of transport companies has sensibly increased in the general context of the global commercial geography. However, the nature of this role is changing as a result of a general reduction of transport costs but growing infrastructure costs, mainly due to greater flows and competition for land. Each transport sector must consider variations in the importance of different transport costs. While operating costs are high for air transport, terminal costs are significant for maritime transport. Several indexes, such as the Baltic Dry Index, have been developed to convey a pricing mechanism useful for planning and decision making. Technological changes and their associated decline in transport costs have weakened the links transport modes and their terminals. There is less emphasis on heavy industries and more importance given to manufacturing and transport services (*e.g.* warehousing and distribution). Indeed, new functions are being grafted to transport activities that are henceforward facilitating **logistics** and manufacturing processes. Relations between terminal operators and carriers have thus become crucial notably in containerized traffic. They are needed to overcome the physical and time constraints of transshipment, notably at ports.

The requirements of international trade gave rise to the development of specialized and intermediary firms providing transport services. These are firms that do not physically transport the goods, but are required to facilitate the grouping, storage and handling of freight as well as the complex paperwork and financial and legal transactions involved in international trade. Examples included freight forwarders, customs brokers, warehousing, insurance agents and banking, etc. Recently, there has been a trend to consolidate these different intermediate functions, and a growing proportion of global trade is now being organized by multi-national corporations that are offering door to door logistics services.

6.2. Economic Efficiency Impacts

A basic tenet of market theory is that economic efficiency is maximized when prices (*perceived variable internal costs*) reflect marginal costs. Mispricing prevents users from efficiently incorporating costs into their consumption decisions." Price is the mechanism by which scarce resources are allocated efficiently between competing uses. For consumers, price encourages a purchase if the benefits of making the purchase exceed the benefits of alternatives. For producers, prices provide incentives fore sources to move to those uses which people value most highly by informing firms how to produce, which products to produce, when and where to sell the products, and when, where and how to invest." Cost estimates in this study indicate that motor vehicle use is significantly underpriced. External costs range from 31% to 42% of total costs, as shown in Table 9.2-1. User costs (both market and non-market) would need to increase 58% to 69% to incorporate all costs.

The under pricing of private motor vehicle travel external costs are not eliminated, they simply show elsewhere in the economy; for example as higher prices for commercial goods (for parking costs), increased local taxes (for road services), higher health care costs (from automobile crashes), illnesses (from pollution and inactivity), and lower residential property values (from urban traffic). Another effect of underpriced driving is that non-automotive travel choices decline. Walking, bicycling, transit, and rail transport receive little capital investment, land use patterns and social habits develop which conflict with these travel options, and they are stigmatized. The majority of automobile costs are either external or fixed, and therefore unrelated to the amount a vehicle is driven. These are forms of under pricing. External costs range from 73% to 113% of user marginal costs (vehicle operating costs plus, user time and accident risk), and 186% to 323% of vehicle operating costs. For each dollar users spent operating a vehicle they incur an average of $2.55 worth of external costs related to congestion, crashes, parking and environmental degradation. Although under pricing driving may appear to benefit users (and indeed benefits many individuals in the short term), mispricing reduces overall economic efficiency. Under priced automobile use increases purchases of transport over other consumer goods, and driving over other travel modes.

Low prices force other constraints, such as congestion, pollution, and resource depletion to limit growth. Under pricing makes these problems unavoidable. Under such circumstances, increasing road capacity is not only ineffective at reducing congestion, it may be counterproductive.

Consumers pay fixed and external costs no matter how much or little they drive, which reduces the incentive to limit driving to trips with positive net value. Put another way, automobile owners receive only a small portion of the savings produce when they driving less. Shifting costs from being external or fixed to being internal and variable increases user choice. At worst consumers would simply shift the money saved in reduced fixed expenses to cover their higher variable expenses, resulting in no overall change in travel or cost. But they can achieve savings that are not currently possible by foregoing low value trips or shifting to more efficient modes. The driving that is eliminated represents existing travel with negative net benefit. Efficient pricing makes consumers better off and improves transportation system efficiency. The total inefficiency and harm of these market distortions is far greater than most analyses indicate because their impacts are cumulative and synergistic (total impacts are greater than the sum of individual impacts). For example, underpriced parking not only increases parking costs, by increasing total vehicle ownership and use it also increases traffic congestion, accidents, pollution emissions and sprawl compared with what would occur with efficient pricing, while under pricing road use, by increasing vehicle trips increases parking costs. Under pricing increases automobile dependency and reduces travel options, which is unfair to non-drivers and reduces transport system efficiency.

Under pricing encourages automobile use for trips when alternatives are more efficient overall. Put differently, correcting pricing distortions provides multiple benefits. Because transport costs are diverse, no single mechanism can capture all externalities. Increased fuel taxes are often recommended to internalize automobile costs,6 but this provides only modest incentive to reduce congestion, parking costs, and some environmental impacts. Others emphasize peak period road pricing, which reduces congestion but not other externalities such as pollution or parking subsidies. Efficient pricing requires several types of fees: weight distance fees, fuel taxes, congestion pricing, smog fees, parking charges, marginalized insurance and higher fines for traffic violators. Transaction costs can make optimal pricing inefficient, which could justify continued under pricing. Current parking charges and traditional road tolls require users to deliver money in the correct form, and impose administration costs. People may prefer to avoid transaction costs by paying lump sums or bundling transportation fees with other charges. New technologies can significantly reduce these transaction costs, lowering the threshold under which marginal pricing is justified. Although transaction costs prevent society from ever achieving pricing that perfectly reflects costs, this is not a barrier to significant progress in marginalizing transport costs. Many costs associated with increasing transportation prices, such as unemployment and reduced profits in motor vehicle dependent industries are *transition* costs which decrease over a few years and eventually disappear. Transition costs are economic inefficiencies that result when durable factors such as location, equipment and training don't match current markets. Transition costs can be minimized by good planning and gradual implementation of changes. For example, the British government's plan to increase fuel taxes 5% annually above inflation for a decade to internalize environmental costs is expected to have minimal transition costs because it is so predictable and gradual.

6.3. Economic Development Implications of Under Pricing

Economic Development refers to progress toward a community's economic goals, including increased productivity, employment, competitiveness, property values, business activity, investment and tax revenues, as summarized below. Transport is a major component of economic activity. In 2005, transportation-related demand accounted for over 10.3 per cent of U.S. Gross Domestic Product (GDP), represented about 10% of total employment, and represented about 18% of household expenditures. As a result, there is a popular assumption that motor vehicle travel is economically beneficial, and constraints on driving (such as increased prices for fuel, road use and parking) are economically harmful. Increased transport prices are claimed to have a "multiplier effect" that raises overall costs and reduces productivity. However, these arguments misrepresent transportation costing impacts. Increasing transportation system efficiency and reducing total transport costs can increase economic productivity, but shifting costs and reducing transportation system efficiency tends to be economically harmful. For example, businesses benefit if their shipping or parking costs decline, but are worse off if underpricing of roads and parking facilities leads to traffic congestion and increases parking demand, increasing their total costs. Although the U.S. produces some motor vehicles and fuel, expenditures on these goods provide relatively little economic benefit because they are capital intensive and mostly imported. Even locally assembled vehicles are built from components largely imported from other regions. There is no evidence that policies which stimulate domestic vehicle and fuel expenditures stimulate economic development overall. A consumer dollar spent

on automobiles means one less dollar for housing, food, entertainment, education, or investment. Consumer expenditures on fuel are particularly harmful to the economy. For example, a Texas case study found that automobile expenditures provide far fewer jobs than expenditures on general consumer goods or public transportation; similarly, shifting expenditures from petroleum and automobiles substantially increases employment. This indicates that policies which reduce consumer expenditures on motor vehicles and fuel tend to increase employment and business activity, particularly in regions that import petroleum.

6.4. The Effects of Underpricing: Two Industry Example

Consider an economy with two industries: Heavy and Light. Transport is a relatively large portion of Heavy's production costs, and a small portion of Light's.

- ☆ These industries initially face a transport price structure which under prices shipping by funding roadways through general taxes. Since both industries pay the same tax, Light is effectively subsidizing Heavy. *Score: Heavy + Light - Economic Efficiency:*

- ☆ A $0.10 per mile road use fee is implemented to replace the property tax, called Fair Price. Heavy industry pays more taxes while Light pays less. *Score: Heavy - Light + Economic Efficiency +*
- ☆ At worst, Heavy pays the full road charge. But, Heavy may change practices to reduce costs such as by moving more goods by rail, so its net cost increase is minimal. *Score: Heavy = Light + Economic Efficiency ++*
- ☆ FairPrice also reduces traffic congestion and accidents. Assume that Heavy industry's shipping time is reduced by 1/2 hour for every 100 miles of trucking and its trucks cost $50 per hour to operate, this means the $10 per hundred mile road user charge saves $25 inoperating expenses, for an overall saving to Heavy. *Score: Heavy + Light + Economic Efficiency +++* Although this example is simplistic, it emphasizes two important points:
- ☆ Underpricing transfers rather than eliminates costs.
- ☆ Full cost pricing provides incentives for business to use resources more efficiently, which provides long term economic benefits, especially to transportation dependent industries. Although a certain amount of motor vehicle travel is efficient and economically beneficial, beyond this optimal, increased driving is economically harmful. A World Bank study indicates that automobile dependency reduces regional economic development. The researchers conclude that, *"there are no obvious gains in economic efficiency from developing car dependence in cities,"* and, *"There are on the other hand significant losses in external costs due to car dependence."* They find that beyond a certain point (about 7,500 kilometers of per capita annual motor vehicle travel), increased driving impose economic costs that outweigh marginal economic benefits.

They suggest that expenditures on motor vehicles and their facilities divert large amounts of private and public capital from alternative uses that are more economically productive. This study finds that the portion of income devoted to commuting is highest in automobile dependent cities, and that transit service efficiency is lowest and vehicle accident rates per capita are highest in automobile dependent cities. Long-term negative economic effects of transport price increases are relatively small, and appear to be declining in most industrial sectors. Freight transport has decreased as a percentage of GNP, industrial production, and national employment due to increased efficiency and more high-value, low-bulk products. Transport accounts for only about 5- 6% of most manufactured product prices. North American fuel prices and user fees are among the lowest among developed nations. Increasing fuel taxes to internalize externalities and reduce demand, with revenues used to replace other taxes, could have would have overall positive economic and employment impacts. Automobile under pricing may have been justified when the automobile industry was first developing, since vehicle and road production experienced economies of scale. This unique historical event does not apply to mature markets. Vehicle use now experiences diseconomies of scale now due to increasing congestion, crash risk and other externalities. Each driver benefits if others drive less, reducing conflicts over road and parking space. There are probably few, if any, further economies of scale in automobile, petroleum and roadway industries.

A review of economic impact research concludes that transportation improvements may provide negative as well as positive economic impacts in a community. Most highway improvements in developed countries provide only marginal productivity benefits. Nijkamp and Blass argue that transport facility investments only contribute to development if other conditions are ripe and transport costs are a significant economic constraint. Kessides concludes that infrastructure investments can provide high returns in developing countries but only normal returns in North America and Western Europe. *Public transit expenditures provide twice the rate of return in productivity as highway expenditures, indicating that increased transport diversity and efficiency provide greater macroeconomic benefits than overall increases in motor vehicle use*. Piet Riveted calls economic arguments for transport subsidies *"scientific mystifications."*

He states that *".the direct contribution of infrastructure improvement to a reduction in transport costs is in general small in industrialized countries."* Harry Dimitriou argues that inter-community transport improvements support economic development, but intercommunity transport improvements (such as urban roadway widening) provide little Economic benefit in developed regions. Some macroeconomic research suggests that transportation infrastructure have high economic returns. However, this does not prove that increased *driving* provides economic benefits. Rather, it implies that increased *transport efficiency* provides benefits. Public transit expenditures provide twice the return on investment as highwayimprovements,30 indicating that it is the general benefits of mobility or accessibility rather than the specific benefits of driving that provides economic benefits.

6.4.1. Land Use Impacts of Underpricing

Transportation has direct and indirect impacts on land values and use. Overinvestment in transportation facilities and under pricing of vehicle use increases the amount of land used for transportation and encourages low-density, urban fringe development. If highways improve access between urban and rural areas, rural property values increase as urban home buyers compete with other rural land uses. This profits current rural land owners if they are willing to sell to developers and allows individuals to purchase more land, privacy, and private environmental amenities than otherwise possible. It also allows exurban residents to compete for urban jobs, and access urban services, benefiting exurban residents. At one time this was considered unquestionably good for society. From society's perspective, all costs and benefits must be considered in each decision, but the perspective of individual decision makers is often more limited. People tend to have different, often conflicting perspectives of transport costs and benefits, depending on their role. These differing perspectives and definitions create conflicts over goals, objectives and strategies, and can result in economically inefficient decisions. Another way to view the conflicts in transportation decision making is to consider the perceived relationships between average costs and quantity of driving from three perspectives: of *users* (who decide how much to travel), *transportation professionals* (who plan transport facilities) and *society* (which bears environmental and social costs and provides facility funding).

6.4.2. Environmental and Social Costs

Since most motor vehicle costs are fixed, marginal costs decline with increased annual milage, giving vehicle owners an incentive to maximize driving. Facility development has a downward sloping cost curve (economies of scale) when traffic is low, since increased driving allows costs to be divided among more miles of use, but once the system is congested average costs increase. Environmental, and social costs of driving are minimal when use is low, but slope upward. Individuals face incentives to maximize driving, "to get their money's worth" from large fixed vehicle costs. The transport agency's U-shaped average cost curve implies economies of scale when roadway development is a goal.39 Only after congestion becomes a problem do transport agencies perceive benefits from reducing traffic demand, and even then they frequently experience incentives to encourage driving, such as fuel taxes dedicated to road building. The upward sloping cost curves associated with other social and environmental costs means that society benefits from reduced driving. As a result of these different price signals the perspectives of individual drivers and automobile oriented transport planners conflict with society in general, and transport professionals who take a broader view of transport costs and choices. The emphasis on increasing automobile capacity, rather than accessibility and other community goals is ingrained in transport planning and financing. The existence of dedicated roadway agencies and funding skews planning decisions toward roadway development.

You may be wondering, *"What does this mean to me?"* Underpriced driving is an economic transfer that allows some households to own more cars, travel more and live in more suburban locations than if prices were higher. But this additional consumption of mobility may provide little net consumer benefit. Under pricing imposes costs that show up elsewhere in household budgets, such as higher consumer prices to pay for "free" parking, higher taxes to pay for roads and traffic services, and increased health costs. At one time under pricing may have provided external benefits due to economies of scale in roadway and industrial development, but there is no evidence that such benefits still exist. The truth is just the opposite. Individuals have no choice but to bear the financial, crash risk and environmental costs of automobile use. Non-drivers have fewer mobility choices. Negative effects of underpriced driving include:

- ☆ *Increased overall transport costs.* Low marginal prices for driving encourage individuals to spend more on driving, leaving less funding for other goods.
- ☆ *Reduced transportation diversity.* Fewer people use alternative modes, so they receive less investment than they otherwise would.
- ☆ *More externalities.* Pollution, congestion, and motor vehicle accidents are surely greater than they would be if user costs were higher.
- ☆ *Increased land devoted to roads and parking facilities.* This means less land available for other uses, higher taxes, or reductions in other government services, and higher consumer costs to pay for "free" parking.

- *Automobile oriented land use, economic and social patterns.* Increased centralization and scale of services and activities, and less emphasis on neighborhood activities, services and relationships. The use and usefulness of streets for non-driving purposes, including walking, playing and other socializing has declined. How these affect any specific individual depends on many factors including driving ability and automobile ownership, income, residence and job location, and future goals. The impacts of transport under pricing are also affected by how much an individual values environmental protection and social equity, and how much they value non motorized transport and community interaction. The benefits of underpriced driving are skewed toward people who drive the most, including those who are relatively wealthy, exurban and rural drivers, and long distance commuters. Children and teenagers, the elderly, the very poor and the handicapped tend to use automobiles relatively little and therefore receive the least benefits of under pricing and suffer significant disbenefits. The effects of price increases depend on how new prices are structured, how quickly and predictably changes occur, whether alternative modes are improved, and how revenues are distributed.

7
Transport Pricing

7.1. Transport Pricing

Transport-pricing policies are aimed at changing car use, and consequently, reducing problems caused by car traffic. Transport pricing may elicit various behavioral changes that may affect collective qualities differently. Therefore, policymakers should base their selection and design of transport-pricing policies on the policy goal at stake and behavioral changes to be reached. In general, transport-pricing policies will be more effective if price increases are significant, and if feasible alternatives are available. Policymakers seem generally reluctant to implement stringent transport-pricing policies, for it is assumed that such policies may significantly reduce individual's quality of life. However, empirical evidence for this assumption is lacking. In fact, studies have revealed that quality of life will not be strongly affected when transport pricing is implemented. One reason for this is that people do not only consider individual consequences of transport pricing, but they also take into account effects on environmental qualities and congestion. Overall, people expect to be better off when pricing policies would reduce environmental and congestion problems. This may well be related to acceptability of such measures, which is confirmed by the positive relationship between perceived effectiveness and acceptability.

Lack of public support is an important barrier for the implementation of transport pricing policies. Acceptability is dependent on various individual perceptions and motivations. First, acceptability is dependent on the expected consequences of transport pricing policies. Transport pricing is more acceptable when people think these policies would actually reduce environmental and congestion problems and when it benefits the individual more (which may be related to improved collective quality). This implies that acceptability may increase when the expected individual and collective benefits are clearly communicated. Process and distributive justice are also loosely related to acceptability judgments; they affect to what extent policies are perceived to be fair. However, little is known yet about the role of fairness, *e.g.*, why are policies perceived to be fair or not, and which justice principles are prevalent for different groups?

Second, acceptability may be related to moral and environmental concerns. Studies in the environmental field suggest that acceptability of policies is higher if people feel a moral obligation to act in the common good. This will be especially the case if they are aware of relevant problems and when they feel responsible for these problems, which are in turn related to general environmental beliefs and values. This implies that acceptability may increase if people are aware of the problems caused by car use, and when they feel responsible to do something about it. Further research is needed to see whether these results may be generalized to traffic and transport issues. Various policy characteristics may be related to acceptability of transport pricing. Obviously, significant price increases will be less acceptable than minor ones. However,

minor price-increases appear to be not acceptable as well because in that case problems resulting from car use will not be reduced while prices of travel increase; as indicated earlier, perceived effectiveness is an important precondition for acceptability. The extent to which transport-pricing policies are differentiated may also affect acceptability. This is closely related to processes of distributive fairness discussed earlier. Fairness of transport policies may be enhanced via revenue allocation. Revenue allocation appeared to strongly related to acceptability judgments. In general, transport-pricing policies are more acceptable if revenues benefit individual car users rather than the general public. Thus, acceptability of transport pricing may be enhanced by clarifying how revenues are used, and by compensating those who are affected most by the policies, as far as this would not decrease effectiveness of policies. Allocating revenues to road infrastructure appeared to be highly acceptable in some studies, but not acceptable at all in other studies. The design of research tasks and research method followed may be the reason for these conflicting results. Further sophisticated research is needed to further clarify the role of revenue use.

Next to technological measures, behavioral changes of individual car users are needed to reduce the problems of car use discussed in the first part of this book. Several policy strategies may be applied to manage car use (see Chapter 15 by Loukopoulos), among which are transport-pricing policies. Prices of car use may be increased (at certain times or places), while costs of the use of environmentally friendly (or: sustainable) modes of transport, such as public transport, may be reduced. Studies revealed that the latter strategy is generally not very effective in the long term. For example, lowering costs of bus use does attract new bus users, but does not result in a reduction in car use, *i.e.*, the new passenger used to cycle or walk, or stayed at home rather than drove their car (*e.g.*, Claassen and Kropman, 1995; Deslauriers and Everett, 1977). Furthermore, financial rewards for bus travellers appeared to be effective only in the short term; people revert to their initial travel behaviour as soon as the reward ceases (Van Knippenberg and Van Knippenberg,

1988). Increasing costs of car use is supposed to be far more effective in reducing the level of car use (*e.g.*, Gomez-Ibanez and Small, 1994; TfL, 2004). Therefore, we focus on price\ increases. In the remaining of this chapter, transport pricing is defined as increasing costs of car use (and not purchase costs of cars). Transport pricing is generally believed to be an effective and efficient way to manage problems resulting from private car use (see Chapter 18 by Ubbels and Verhoef). Especially modelling studies, but also some evaluation studies, revealed that transport-pricing policies may be highly effective in reducing car use and the problems associated with car use.

However, transport pricing is not easily implemented due to the lack of public support, *i.e.*, in general, the public thinks transport-pricing measures are not acceptable (*e.g.*, Schlag and Teubel, 1997; Jones, 1998, 2003). This also applies to other policies: effective policies are generally not acceptable to the public, while policies that are acceptable are generallynot effective in reducing problems resulting from car use (Steg, 2003). This places policymakers in a dilemma: Should they implement unpopular policies, or respect public opinions without solving the problems at stake? Or is it possible to design pricing schemes that are both effective and acceptable? Therefore, it is important to understand which factors affect the effectiveness and acceptability of transport policies, and more specifically, transport pricing. This facilitates the design of policies that may be acceptable and effective. Characteristics of policies (*e.g.*, price level) as well as individual characteristics (*e.g.*, car dependency, car attitudes) may be important in this respect.

This chapter provides a psychological perspective on transport pricing. We first discuss whether transport pricing is indeed effective in reducing car use. We will briefly review studies that have examined the effectiveness of transport pricing. Some of these studies focus on perceived effectiveness (*i.e.*, stated preference), whereas others evaluated actual effectiveness (*i.e.*, revealed preferences). Next, we focus on acceptability of transport-pricing policies, since acceptability is of major importance for implementing effective transport-pricing policies. We discuss factors influencing the acceptability of transport pricing and provide theoretical explanations for the significance of these factors. Furthermore, we elaborate on the relationships between acceptability and effectiveness judgments: Are effective policies indeed not acceptable to the public? Several policy recommendations are discussed, and important open questions that need to be addressed in future research are highlighted.

7.2. Effectiveness of Transport Pricing

Various transport-pricing policies may be distinguished, such as fuel levies, road taxes, toll levies, and kilometers changes. Type of Change Specification of Behavioural Changes:

- ☆ Driving behaviour Driving style (speed, use of brakes, changing gears)
- ☆ Travel behaviour Trip chaining
- ☆ Route choice
- ☆ Time of travel
- ☆ Destination choice (*e.g.*, shopping, recreation)

- Number of trips (trip suppression)
- Mode choice (*e.g.*, car, carpool, motorbike, public transport, bicycle, walking)
- Vehicle ownership Type of car (*e.g.*, fuel type, size)
- Car ownership (number of cars, replacement of car)
- Location choice of residence
- Choice of workplace

One may decide to drive more often or to drive during peak hours instead of off-peak hours if congestion levels reduce. Similarly, a reduction in congestion problems may result in an increase of commuting distances for some groups, *e.g.*, when people can afford to move residence further away from work without increasing their travel time. Changes in car ownership and location choices (especially choice of workplace and residence) will typically occur in the long term. One may assume an ordering in these adaptation strategies (see also Loukopoulos *et al.*, 2004). People may first opt for behavioral changes with lowest behavioral costs (in terms of money, time, effort), *e.g.*, trip chaining (*e.g.*, combine commuting and shopping trips), choosing closer destinations (*e.g.*, shopping close to home), or choosing different routes (*e.g.*, travel via a route that is not tolled). When these changes are not possible or not sufficient, they may opt for changes associated with higher behavioral costs, such as using other modes of transport or suppressing trips.

Finally, in the long term, they may consider buying another car or disposing of a car, or even to move home or to look for a job closer to home. Of course, individual differences may exist in behavioural costs of the adaptation strategies. For example, changing the time of travel may be possible for some workers but not for others. To get a complete and comprehensive overview of effectiveness of transport pricing, it is highly important to distinguish between these types of behavioural changes and to take into account short as well as long-term effects of transport pricing. People need time to adjust to new situations, and more significant behavioural adaptations may occur in the long term only. For example, Mogridge (1978) found that during the world oil crisis of the mid- 1970s the rise in fuel prices had only a marginal effect on car use in the short term. However, it appeared that behavioural effects were apparent in the long term: people purchased smaller and more economical cars. More generally, price elasticities tend to be higher in the long term compared to short term (*e.g.*, Priemus and Nijkamp, 1994; Wootton, 1999). Different pricing strategies may result in different behavioural adaptations. For example, a rise in vehicle or road taxes will probably affect car ownership more strongly than route choice, while a kilometer charge may well result in changes in mode choice, but probablynnot in the purchase of a more energy–efficient car. Furthermore, time-dependent charges are more likely to result in changes in time of travel, while tolling main roads may typically result in changes in route choice.

The extent to which the behavioural changes discussed above will contribute to reduction in the various problems related to car use differs greatly. Changes in route or time of travel may be effective in reducing accessibility problems, but may be counterproductive in attempts to increase quality of urban life. For example, if heavy traffic volumes are spread out in time, community residents may suffer from high levels of traffic noise for a longer period of time. On the other hand, reductions in the level of car use, *e.g.*, by shifting to less polluting modes of transport, changing destinations, residence or workplace, combining trips, or suppressing trips may improve environmental quality, urban quality of life, and destination accessibility. Purchasing new, energy–efficient or low-emission cars will especially result in reduction of environmental problems and increases in urban quality of life, but will hardly help control congestion problems. People may be even tempted to use their energy–efficient car more often because it is cheaper and more environmentally friendly. Many scholars have stressed the importance of the so-called rebound effect (*e.g.*, Berkhout *et al.*, 2000). Therefore, policymakers should carefully take into account which policy goals should be reached and which behaviour changes are targeted when designing and implementing transport-pricing policies.

Next to changes in collective qualities, such as environmental problems, urban quality of life and accessibility, transport-pricing policies may also have consequences for individual quality of life (*e.g.*, Poortinga *et al.*, 2004; Steg and Gifford, 2005; De Groot and Steg, in press). Quality of life may be defined as the extent to which important values needs of people are fulfilled (*e.g.*, Diener, 1995; Diener *et al.*, 1999). Obviously, transport-pricing policies may affect the way and the extent to which people are able to fulfil important values and needs. In general, policymakers seem reluctant to implement policies aimed at reducing car use, such as increasing prices of car use, because it is believed that such measures will reduce individual's quality of life. However, various studies reveal that transport policies may have less negative effects for individual quality of life than expected (De Groot and Steg, in press; see Steg and Gifford, 2005, for a review). Possible negative quality-of-life effects, such as infringements on freedom, reductions in comfort and increased costs, seem to be partly compensated for by improvements in environmental quality, nature, health, and safety, which are considered to be important for individual quality of life as well. A study by Schuitema and Steg (2005b) also revealed that people take into account effects on environmental quality and congestion when evaluating effects of transport-pricing policies for their lives. People's judgements of whether they would be better or worse off when considering all pros and cons of transport-pricing policies appeared to be most strongly (and positively) related to expected effects on congestion levels and environmental problems, while these judgements were only weakly correlated with the extent to which the pricing policy would affect one's own car use (*i.e.*, car mileage). This implies that respondents expect their overall situation to improve if they think congestion and environmental problems will decrease as a result of transport-pricing policies (see also Schuitema and Steg, 2005a).

Psychologists have primarily studied the expected effects of transport pricing (*i.e.*, they focus on stated preferences). Typically, respondents are asked to what extent they would reduce their car use (*e.g.*, mileage, number of trips) when transport-pricing policies would be implemented, or to what extent these policies would be effective in reducing problems caused by car use. Various studies revealed that, in general, people think transport pricing will not be very effective in reducing car use (*e.g.*, Steg, 1996, 2003; Jakobsson *et al.*, 2000; Loukopoulos *et al.*, 2004; Schuitema and Steg, 2005b) or the problems of car use (*e.g.*, Schlag and Teubel, 1997; Jones, 2003; Schuitema and Steg, 2005a, 2005b), especially because they think car use reductions are hardly feasible (Jakobsson *et al.*, 2002). People generally assume transport pricing will be somewhat more effective for others (*e.g.*, Steg, 1996). Various group differences were found in perceived effectiveness of transport pricing. In general, low-income groups expect to reduce their car use more than high-income groups do (Jakobsson *et al.*, 2000). Furthermore, transport pricing will affect private and commuter trips somewhat more than business trips (Cavalini *et al.*, 1996; Schuitema *et al.*, 2003). Transport pricing may especially result in a reduction of shopping trips (Jakobsson *et al.*, 2002; Schuitema *et al.*, 2003).

Furthermore, transport policies are more likely to result in changes in route choice, destination choice or time of travel, while the amount of kilometres driven or car ownership will hardly be affected (Cavalini *et al.*, 1996; Loukopoulos *et al.*, 2004). Especially, short trips by car will be reduced, *e.g.*, because people cycle to their destination instead of driving (Schuitema *et al.*, 2003). In sum, stated preference studies reveal that transport pricing will be more effective if people think they can reduce their car use (*e.g.*, this is more likely for private trips and route choice), and if people cannot afford an increase in travel costs (*e.g.*, low-income groups). Surprisingly, people indicate that in general, transport pricing will not be very effective, while various studies reveal that transport pricing may have substantial effects on car use. Prominent examples are the Singapore area licence scheme (a congestion charge) and the London congestion charge (see Small and Gomez-Ibañez, 1998; Santos, 2004; Santos *et al.*, 2004; Verhoef *et al.*, 2004). In Singapore, changes in route choice were most prevalent, but shifts to carpooling and bus use were observed as well, while in London switches to bus use (and taxi's) were more common than changes in route choice (see Santos, 2004; Verhoef *et al.*, 2004). Thus, revealed preferences do not seem to match stated preferences. Several explanations may be given for this finding. First, results of revealed preferences studies may not be easily generalised to other areas.

Congestion pricing appeared to be effective in Singapore and London, which both were confronted with serious congestion problems. Moreover, in both cities, the quality of public transport was improved as well, and consequently, for many, feasible alternatives modes of transport were available. Experiencing serious problems and the availability of high-quality public transport may be important preconditions for transport pricing to be effective. Also, the level of price increase is important. In both Singapore and London, charges were relatively high. In contrast, low charges appeared to be less effective (*e.g.*, toll rings in Norway; Tretvik, 2003; Ramjerdi *et al.*, 2004; see also Verhoef *et al.*, 2004). Also, the effectiveness of transport pricing may be substantially decreased when people can evade the measure, *e.g.*, by driving at other times when time-dependent charges are introduced (such as in Trondheim, Norway, time-dependent urban tolls were implemented; Meland, 1995; Ramjerdi *et al.*, 2004). Furthermore, stated preference methods may elicit strategic answers; people indicate transport pricing will not be effective, in the hope it will convince policymakers to not to implement transport-pricing policies.

Moreover, people may not think through the consequences of transport-pricing policies, and do not thoroughly consider possible behaviour changes. In general, people are not very accurate in assessing future travel patterns (cf. Jakobsson, 2004) Pricing policies may have unwanted side effects, resulting in less significant behaviour changes than expected beforehand. This will especially be the case if pricing policies reduce intrinsic motivation to help solve problems related to car use. This so called 'crowding out' effect (*e.g.*, Frey, 1997, 2003) may especially occur when price changes are high enough to justify behavioural changes. In that case, people may change their car use to save money only, and no longer because they feel intrinsically motivated to contribute to solutions of the problems caused by car use. This may be problematic because various studies have revealed that intrinsic motivation may more strongly affect behaviour than do monetary incentives (Thøgersen, 2003). Moreover, intrinsic motivation of those not affected by the pricing policies may diminish as well. Thus, pricing policies may be effective only to the extent that financial incentives at least compensate a possible decrease in intrinsic motivation (Frey, 1997; Thøgersen, 2003). In other cases, effects may be weakened, or may even be counter-productive. Transport pricing may give people the feeling that they have a right to drive their car because they have paid for it (Fehr and Falk, 2002). To the authors' knowledge, it has not been demonstrated yet whether the crowding out effect applies to transport pricing, but based on results of previous studies, such effects may be expect.

7.3. Transport Pricing

Lack of public support is an important barrier to the implementation of restrictive transport-pricing policies. Interestingly, policymakers and politicians seem to underestimate the support for policies aimed to reduce car use (*e.g.*, Jones, 1995). A similar misunderstanding is prevalent among car users themselves, who, on average, indicate they find transport pricing more acceptable than does the general public (Steg, 1996). Nevertheless, many politicians are not keen on implementing policies that evoke resistance among the general population. However, some transport-pricing policies are more acceptable than others, and some may even be rather acceptable, at least for some groups. In general, the extent to which transport pricing is acceptable to the public is dependent on individual characteristics as well as on features of the policies themselves. In this section, we will discuss both types of features.

Many (psychological) studies have tried to identify individual factors that affect public acceptability of transport pricing (*e.g.*, Schlag and Teubel, 1997; Jakobsson *et al.*, 2000; Schade and Schlag, 2000, 2003; Schlag and Schade, 2000; Bamberg and Rölle,2003; Schuitema, 2003; Loukopoulos *et al.*, 2005). The acceptability of transport pricing appears to be related to a wide range of factors. A study by Frey (2003) suggests that problem awareness may be an important precondition for policy acceptability. Acceptability of road pricing in Saas-Fee (Switzerland) appeared to be rather high (57% supported this policy) because people were aware of the problems caused by car raffic and of possible solutions for these problems in their community. Similar results were found in a study conducted in the Netherlands: transport policies are more acceptable for people high in problem awareness (Steg and Vlek, 1997). Jones (2003) also stresses the importance of problem awareness for enhancing public acceptability for transport pricing (see also Small and Gomez-Ibañez, 1998). Interestingly, Schadeand Schlag (2000) found that people who are especially concerned about the environmental problems of car use evaluate transport-pricing policies as more acceptable compared to those who are more strongly concerned about congestion.

Jakobsson *et al.* (2000) found that transport policies are evaluated as less acceptable when they are perceived to be unfair, and when they threaten people's freedom of choice (and thus have negative individual consequences). A study by Bamberg and Rölle (2003) revealed that, besides these factors, perceived effectiveness plays a role as well: acceptability is higher when people believe the policies will be effective in reducing car use (see also Steg, 1996; Schade and Schlag, 2003). Perceived effectiveness appeared to be related to problem awareness: the more people are aware of environmental problems caused by car traffic, the more they think transport pricing will be an effective instrument to reduce car use. Several other studies reported positive relationships between perceived effectiveness and acceptability of transport pricing as well. Jaensirisak *et al.* (2003) found that acceptability is higher if transport pricing benefits individuals as well as society as a whole by, *e.g.*, reducing congestion levels and improving environmental quality. Schade and Schlag (2003) also report that transport pricing is more acceptable if it benefits the individual and reduces collective problems. Similar results were found by Schuitema and Steg (2005a), showing that acceptability of transport-pricing measures is higher if people think their life will not be affected too much, and if people think environmental and congestion problems will actually reduce. They found that the actual risein travel cost for households was less strongly related to acceptability ratings.

People who think transport pricing is acceptable especially stress the positive (collective) consequences of the policies, such as improved environmental quality and reduction in congestion levels, while those who think transport policies are not acceptable typically focus on negative consequences for themselves (Loukopoulos *et al.*, 2004). From the above, we may conclude that people seem to resist policies that are not effective in solving problems caused by car use. On the other hand, Jakobsson *et al.* (2000) found that policies are not acceptable when they are very effective in changing one's behaviour, thereby seriously affecting individual freedom of choice. This implies that people seem to prefer policies that help solve collective problems resulting from car use that benefit themselves as well, while the policies should not restrict their own freedom of choice. Various groups may actually benefit from the implementation of transport-pricing policies, for example if congestion levels decrease and accessibility improves. Furthermore, transport-pricing may benefit groups that hardly drive or do not drive, as far as the problems resulting from car use actually reduce, *i.e.*, they will benefit from improved environmental and urban qualities. Also, people may find ways to evade the charges, and consequently, price increases may not affect them. Thus, some groups may be in favour of transport pricing. However, in general, support of these 'winners' is usually muted (Frey, 2003), and, consequently, politicians assume public support is low.

Trust in governments may also affect policy acceptability: if trust levels are low, public support for policies that reduce individual freedom to move is low as well (Ney *et al.*, 1997). This may well be related to perceived effectiveness: if people do not trust the government, they may think the policies may not be effective as well. Frey (2003) suggests that public participation in the decision-making processes may enhance public acceptability, because it increases public understanding of the need for reducing problems related to car use by introducing transport pricing. This is in line with literature on\ procedural justice, in which the focus is placed on the fairness of the processes by which decisions and policies are being made: decisions are perceived to be more fair, and consequently, acceptable, if those affected had a say in the decision-making process (Lind and Tyler, 1988; Clayton, 2000). Processes of distributive justice, *i.e.*, the extent to which various groups are affected by transport-pricing policies (see Clayton, 2000; Schroeder *et al.*, 2003), are also highly relevant for the acceptability of transport pricing. Indeed, fairness seems to be strongly related to acceptability judgements (Jakobsson *et al.*, 2000; Bamberg and Rölle, 2003;Schuitema *et al.*, 2004). For example, Bamberg and Rölle (2003) report a correlation coefficient of 0.80 between perceived fairness and acceptability. This suggests that perceived fairness is of crucial importance of policy acceptability. On the other hand, one may doubt whether respondents interpret fairness and acceptability as different constructs; they may reflect one common underlying construct. This indicates that measures of both constructs should be carefully developed, as to make sure respondents notice the conceptual difference.

Surprisingly, little is known yet about perceived fairness of different types of transport pricing. What is perceived to be fair will likely depend on the prevalent justice principle?Different justice principles may be prevalent, resulting in different fairness judgements. People may strive for equality or equity, or they may think the 'polluter should pay'. Equality refers to aiming for equal rights, privileges, quantities, qualities, etc. This implies that transport price increases should be the same for all (*i.e.*, static pricing measures). Equity means everyone should be affected in the same degree. This suggests that different price levels for different groups may be evaluated as fair, *e.g.*, price level should be lower for frequent travellers (*e.g.*, business travellers) as to make total price increases equitable, or people driving large and heavy cars (more often high-income groups) should pay higher prices because they can afford it. 'The polluter pay' principle implies that price increases should be highest for those who most strongly contribute to the problems.

Congestion charging is a prominent example here. Fairness judgments may also be directly related to the extent to which transport policies affect the individual relative to other people: if people think that they themselves are affected disproportionately, policies may perceived to be unfair. Another (distributive) justice principle that proved to be relevant in the environmental context is environmental justice, which respects the rights of the environment, ecosystems, other species, and future generations (Clayton, 2000). Policies may be perceived as more fair when they protect and guarantee the rights of nature, the environment, and future generations. Environmental justice may be well seen as a special type of distributive justice, by taking into account the extent to which non human species and future generations are affected by policies. Little is known yet on which justice principle is prevalent for which groups and under which circumstances. Studies typically include quite general measures of fairness, reflecting how fair the introduction of the measure would be (often a single-item measure). More research is certainly warranted into the perceived fairness of various transport-pricing measures and the role of various justice principles in this respect.

This approach makes sense when considering acceptability judgements as attitudes towards transport pricing (*e.g.*, Jakobsson *et al.*, 2000; Schade and Schlag, 2003), *i.e.*,many theorists assume attitudes are based on evaluating perceived consequences ofbehaviour or situations (*e.g.*, Fishbein and Ajzen, 1975; Ajzen, 1985). To put it differently, attitudes towards transport pricing are believed to be dependent on expected consequences of transport pricing. The more positive (and the less negative) consequences people expect from transport pricing, the more favourable are people's attitudes towards transport pricing, and the more positive people's acceptability ratings. Alternatively,acceptability judgements may be defined as a specific type of (environmental)behaviour. This approach has been advanced by Stern and his colleagues (Stern *et al.*, 1999; Stern, 2000). They argue that support or acceptability of environmental policies,such as transport pricing, may be defined as non-activist behaviours in the public sphere. Non-activist behaviour affects environmental quality indirectly, by influencing publicpolicies, which may have large effects on environmental qualities because public policies may change the behaviour of many people at once. This implies that theories aimed to explain environmental behaviour may be of particular interest to understand factors affecting public acceptability of transport pricing. Since transport pricing generally implies that people have to give up personal advantages of car use to benefit collective interests, especially theories that explain why people make short-term sacrifices in order safeguard collective interests seem relevant in this respect, such as the norm-activation model (NAM) (Schwartz, 1997; Schwartz and Howard, 1981) and the value-belief-norm theory of environmentalism (VBN theory) (Stern, 2000). The causal chain proposed in the VBN theory moves from relatively stableand general values, to beliefs about human–environment relations, which in turn arebelieved to affect specific beliefs on consequences of environmental behaviour and the individual's responsibility for these problems and for taking corrective actions, which affect personal norms, and, subsequently, behaviour.

7.4. Policy Features that Affect Acceptability of Transport Pricing

In this section, we discuss policy features that may affect acceptability of transportation pricing policies, and explain the relation to individual factors influencing acceptability judgements as identified in the previous section, notably perceived effectiveness of policies, the extent to which individual freedom to move is affected, and perceived fairness of policies. Three policy features are discussed: (1) the level of price changes, (2) the extent to which price changes are differentiated, *e.g.*, in time, place, for user groups, and (3) the way revenues are allocated. Obviously, the level of price changes affects people's freedom to move. Marginal price increases will hardly affect people or may even not be noticed, while many people (especially low-income groups) may not be able to cope with extensive price increases. Not surprisingly, significant price increases are evaluated as less acceptable than are small increases (Schuitema *et al.*, 2003), especially if the policy is not differentiated and people cannot easily evade the measure. Moreover, significant price increases are not believed to be fair, because low-income groups will be afflicted disproportionately. On the other hand, small price increases may not be acceptable as well (Rienstra *et al.*, 1999; Schade and Schlag, 2000), because in that case travel costs will increase without problems resulting from car use being reduced. This implies that price increases are most acceptable when they are sufficiently high to reduce problems resulting from car use, without affecting one's own preferred travel behaviour.

Furthermore, acceptability may depend on the extent to which pricing policies are differentiated. First, pricing measures may be static, *i.e.*, when price increases are similar for all drivers. Examples are flat kilometre or toll charges. Second, price increases may be variable, *i.e.*, they may differ in time (*e.g.*, higher tolls during peak hours) or place (*e.g.*, higher price levels at specific routes, such as cordon charging), according to the level of the problem (*e.g.*, congestion charging) or vehicle type (*e.g.*, higher charges for large and heavy cars), and for different user groups (*e.g.*, different rates for private cars and lorries). Third, dynamic price increases may be implemented, *i.e.*, price levels may change in time or place, according to actual travel conditions (*e.g.*, congestion levels). In case of dynamic pricing policies, at a given time actual price levels may be uncertain. As people prefer certain outcomes above uncertain ones (Kahneman and Tversky, 1984), we may hypothesise that dynamic pricing is less acceptable. In practice, however, actual price levels under dynamic pricing may be quite stable in as far as equilibrium in travel flows is reached. Some types of differentiations may be fair than others, because they affect various groups differently. For example, congestion pricing will affect drivers in congested areas only, and higher tolls during peak hours will strongly affect commuters. Thus, freedom to move of some groups may be seriously affected, while other groups may not be confronted with price increases at all. In as far as people think some groups are affected disproportionately, they may think the policy is unfair.

Finally, revenue allocation may affect acceptability of pricing policies. Several studies revealed the way revenues are used may have important consequences for acceptability of transport pricing. In general, price increases appear to be more acceptable when revenues are spent in a way that benefit the individual directly, *e.g.*, by reducing taxes for owning or using a car (*e.g.*, Jones, 1991; Verhoef, 1996; Schade and Schlag, 2000).

8

Transport in Nigeria Development

8.1. Transport in Nigeria Development

Transport plays a key role in the economic and social development of any nation. A well functioning and integrated Transport System amongst other things:

- ☆ Stimulates national development and enhances the quality of life for all;
- ☆ Allows markets to operate by enabling the seamless movement of goods and people;
- ☆ Provides vital links between spatially separated facilities and enables social contact and interaction;
- ☆ Provides access to employment, health, education and services;
- ☆ Alleviates regional inequality and fosters national integration;
- ☆ Increases access to markets and links local, regional, national and international markets; and
- ☆ Promotes economic development by increasing access to labour and physical resources thus facilitating the realization of a country's, comparative advantages.

The different Transport modes (road, rail, maritime, air and waterways) have specific features and advantages. The overall efficiency and effectiveness of the Transport system depends on the development of these modes and their interfacing/ integration. The evolution of modern transport system in Nigeria can be categorized into two distinct phases. These are:

- The colonial period this marked the origin of modern transport system. The networks of rail, water and road developed then were geared essentially to meet the exportation of cash crops, such as groundnuts, cocoa, cotton and palm products and to the importation of cheap, mass produced consumption goods. This early transport system were planned in the most economic way possible, as typified in sub-standard road and rail alignments and a sub base, which later proved inadequate to accommodate heavy vehicles.
- The post colonial period/attainment of independence. With a re-orientation of goals, transport became one of the instruments of unification of the country and an important tool for social and economic development. The development of petroleum resources from the 1950's had significant impact on the nation's social and economic growth, putting increasing demands on the transport system. Goods and passenger movements in Nigeria are performed mainly by road, with the railway and inland waterways playing significant, but less important roles. International freight movement is principally by sea while air transportation is the main passenger carrier.

Government's investment in the country's transport system is substantial. The replacement value of road network at 2001 prices is estimated at between N3,500 – N4,300 billion. The fixed assets of Nigerian Ports Authority amounted to N36 billion in 1999, but this represents an addition of unadjusted historical data. The replacement value could therefore be much higher. Investments in the Nigeria Railway Corporation are smaller, approximately N9.5 – N10 billion. All the transport sub-sectors suffer from the effects of past shortages of resources, and this inadequacy has consistently been reflected in inadequate maintenance. In the road transport sub-sector, the lack of road maintenance often necessitates subsequent reconstruction. The inadequate replacement and the poor maintenance of vehicles, contribute to high social costs of atmospheric pollution, and results in high operating costs. In the case of the railways, the lack of necessary resources to keep tracks, rolling stocks and maintenance facility in reasonable working condition produced a serious deterioration of the railway system. Similar problems affect the inland waterways, resulting in its inability to perform useful functions. The long-term shortages are growing, becoming seemingly insoluble. For transport operators, these problems create a feeling of despondency, decline in morale and exacerbate management/1employee tension. Feeling of despondency and low morale cause further deterioration and growing problems of the system, creating a vicious circle. Breaking this circle is the overriding immediate objective of the country's national transport policy.

8.2. Challenges and Opportunities

The Nigerian transport system faces great challenges and also offers several opportunities. The predisposing factors include:

- The size of the country: With area coverage of 923,768.64 km2, spanning longitude 30E-160E and latitude 40N-140N Nigeria is a comparatively large country. It is inhabited by over 140 million peole. Transportation is a crucial instrument for linking the country economically, socially and politically.

- Nigeria's vast natural resources: Nigeria is endowed with petroleum, gas and solid mineral resources. The country is also blessed with abundant agricultural and forest resources, which vary by the three ecological zones of the country. The country is endowed with a highly productive open sea with abundant and diverse marine resources within her coastline of 852km bordering the Atlantic Ocean in the Gulf of Guinea and her maritime area of 46,000 km. This variation is a factor in spatial interdependence. Transport plays an important role in the exploitation and distribution of these resources and in the reduction of spatial inequality and in poverty alleviation.
- The Growth of the Nigerian Economy: Until the current decade, economic growth posed significant challenges to the Nigerian economy, especially during the decades of the 1980s and 1990s. However, between 1999–2008, the performance of the Nigerian economy improved significantly. Real Gross Domestic Product (GDP) grew at an annual average rate of 5.6% and was the highest in three decades. The economic growth rate of almost two times the estimated 3% growth rate of the population ensured a real per capita output growth of 2.6%. The oil sector tended to constitute a drag on growth. The composition of national output for the 1999 – 2008 decade shows that there is still the challenge of achieving broad-based growth. Three large sectors – crop production, oil and gas, and wholesale and retail trade contributed the most to GDP. The role of Manufacturing was limited while wholesale and retail trade, through a high contributor to GDP, had low linkage with domestic production due to low domestic production output and high importation. Overall, the Sectors of the economy are still plagued by numerous problems notably, poor and decaying infrastructure, epileptic power supply, institutions and regulatory deficit, etc. The defective structure of the nation's economy make her heavily dependent on export earnings from crude oil and on import of consumer and capital goods, thereby increasing the country's cargo throughput and boosting the overall maritime trade. The huge volume of cargo generated by Nigeria places her at an economic advantage in South-Saharan Africa.
- Strategic Location of Nigeria: by its location, Nigeria has an extensive coast washed by the Atlantic Ocean and a geographically conducive shore which provide the opportunity of port services to landlocked countries in West Africa and particularly Chad and Niger Republic, as well as serve as a hub for transhipment for the West and Central African sub-region. Thus, the demand for transport is high and growing. This National Transport Policy acknowledges the need to support the country's social and economic growth to enable her compete favourably in international markets. This Policy above all, acknowledges the need to develop a transport system that ensures national security, opens up inaccessible areas and serves the overall well being of the people.

8.3. The Need for a Nigeria Transport Policy

The introduction to the National Transport Policy document of 1993 states that: "At present, the Nigerian transport system functions in a crisis situation", and one of the principal causes, it identified was "a major imbalance between the needs of Nigerian society and economy for adequate transport facilities and the ability of the transport sector to meet such demands". This statement remains valid today, in 2010, in respect of most of the transport system. The imbalance in supply and demand for transport capacity overall, and in the development of the different modes of transport, has in fact increased over the period since 1993. This does not mean that during these years no efforts were made to improve and maintain the system and make it function. A significant improvement exists, but overall the demand for transport services in Nigeria seems to exceed the supply. The Nigerian transport system is still in a very difficult situation that needs urgent remedies. Changes and improvement are obviously needed. Therefore, the present National Transport Policy strives to attain maximum realism both in the identification of the problems and in the assessment of the means to rectify them. The responsibility for planning, developing and maintaining the nation's transport infrastructure is shared among the three tiers of Government. To this end, intra-state roads are the responsibility of State Governments, while the Local Governments are required to cater for intra-urban and rural feeder roads, which account for over 60% of the existing road network. The Federal Government is responsible for the national highways which constitute only 17% of the existing road network. In addition, the Federal Government through its Agencies is also responsible for Inland Waterways/ River Ports, Sea Ports, railways, airports and pipelines.

Furthermore, high-level policies of government in particular, the NV 20:2020 economic transformation blueprint, 7-point Agenda, NEEDS, Public Private Partnership (PPP) and International Commitments such as the Millennium Development Goals (MDGs) all require a functional, reliable and effective transport system to, among other things, connect people, places, services, opportunities, etc. This National Transport Policy therefore provides the guidelines for planning, development, co-ordination, management, supervision and regulation of the transport sector as well as:

- ☆ Explaining Government's decisions and actions in the sector by espousing the goals and principles that guide it;
- ☆ Identifying existing gaps and short-comings and how to address them;
- ☆ Showing how actions in the different modes are linked in pursuit of common goals;
- ☆ Providing the basis for a system of monitoring and accountability; and
- ☆ Ensuring consistency in the application of policy principles across all modes and in pursuit of different objectives.

The fundamental goal of this Nigeria Transport Policy is to develop an adequate, safe, environmentally sound, efficient and affordable integrated transport system within the framework of a progressive and competitive market economy. An "adequate transport system" means that available transport infrastructure and services meet the needs of all Nigerians for access to the market, place of employment and to basic social services. The transport system will serve as an instrument for social, political and economic unification; strengthening the operation of markets, facilitating production and resource development, and promoting relationships with the outside world.

A "Safe transport system" means that all reasonable standards are set and actions taken to prevent accidents and minimise the number of potential victims and the destruction of property. Effective safety measures should protect transport operators and their employees, users of transport services and the public at large. An "environmentally sound transport system" means that reasonable, effective actions will be taken to diminish atmospheric, water and other pollution, through proper planning of infrastructure and the establishment of appropriate regulatory standards. An "efficient transport system" means that the transport services are provided in a way that ensures resources are used efficiently and the economic potential of appropriate technology is used to achieve sustainable gains in productivity in order to reduce costs and improve service quality. An efficient transport system also implies the progressive reduction of social costs, the control of other external costs of transport, and the expenditure of public funds in a way that is properly justified and carefully managed.

An "affordable transport system" means that adequate transport services can be enjoyed by all classes of Nigerians at reasonable cost, and where market mechanism fails to provide this, the Government will intervene to support the provision of essential transport services. An "integrated transport system" means the effective connectivity between ports, rail, road, inland waterways and air, thereby making use of the advantages of different modes to ensure seamless movement of goods and people and better utilisation of resources. Given the current condition of much of the transport sector, the rate at which the goals can be achieved will be subject to the realities of resource availability and practical implementation constraints. It will take some considerable time to make the scale of improvements needed across each of the transport modes to achieve these policy goals. It is therefore essential that the implementation of policy is based on a careful prioritisation of goals and objectives and that plans in each of the transport modes are realistic and consistent.

All transport investment should be subject to careful analysis to ensure that the benefits, net of the costs, meets appropriate investments criteria and represent value for money; and that policies and plans are practical and sustainable. The Government's policy objectives for the sector are:

- ☆ To promote economic development, expand trade, and improve Nigeria's competitiveness through an efficient and affordable integrated transport system;
- ☆ To encourage and remove all barriers towards the private sector participation in the development, provision, maintenance, operation, and upgrading of transport infrastructure and services;
- ☆ To promote the use of public transport over private cars.

- ✰ To promote a culture of maintenance and continuous upgrading of transport infrastructure and services;
- ✰ To promote competition and efficiency and cost reduction of transport services in Nigeria.
- ✰ To improve the safety, security, reliability, quality, and speed of movement of goods and people, at local, national, and international levels;
- ✰ To develop transport infrastructure that ensures environmental sustainability and internationally accepted standards; and
- ✰ To support States and the Federal Capital Territory in the development and promotion of urban transport systems and local governments in developing and promoting rural accessibility.

9

Pricing in Transport: A Multimodal Perspective

9.1. Pricing in Transport: A Multimodal Perspective

Pricing plays a major role in transport policies in many countries. The well-known motivations of pricing policies are the promotion of efficiency and equity. Efficiency leads to the rule that prices are based on marginal social costs implying that the marginal benefits of transport activities equal the marginal social costs. These social costs usually depend on four elements:

- ☆ Costs related to time devoted to transport,
- ☆ Costs of inputs acquired via the private market (for example with car use),
- ☆ Costs related to the services rendered by the public sector (for example most of infrastructure and
- ☆ External costs imposed on others (pollution, congestion).

There are several main problems related to the application of marginal cost based pricing (see for a discussion also Verhoef, 1996, Rothengatter, 2003 and Nash 2003). Below we give a short description.

(a) Measurement Difficulties

The measurement of some cost components mentioned above is not straightforward. In particular the valuation of external costs of transport is a field where still many uncertainties exist. In a number of fields such as transport safety and noise, considerable progress has been made during the past decades, but in other fields, such as the appropriate valuation of CO_2 emissions, and the intrusion effects of transport there is still much uncertainty.

(b) Financing Problems

The use of marginal costs as a principle for pricing may well lead to financing problems for the public sector. For example, most transport infrastructures are uncongested, implying that charging users for the marginal costs would not be sufficient to cover the total costs. Hence other sources of finance would be needed.

(c) Imperfections in other Markets

Given the key role of transport as a link between economic sectors and also given the interdependencies between transport sectors, the simple use of marginal cost pricing in a transport sector is not necessarily efficient. For example, charging one transport modefor NOx emissions, while ignoring emissions in other modes may lead to a modal shift so that only a small part of the potential welfare gain is achieved. This is an example of market imperfections. These imperfections call for second best pricing strategies where behavioral responses in imperfect markets are anticipated. This obviously adds to the complexity of pricing strategies, since no longer is it sufficient that the marginal costs should be correctly estimated for one particular transport mode, but also the marginal costs in other transport modes should be estimated, and on top of that also the effects of pricing measures on modal shift should be considered.

(d) Implementation Costs

Implementation costs of marginal social cost pricing may be high. This depends strongly on the type of cost considered. For example, fuel taxes are easy to implement and are an appropriate tool to address CO_2 emissions. But on the other hand, congestion based charging strategies may lead to rather expensive charging systems. The good news is that with the present trends in information and communication technologies the prospects for cost reductions are favourable.

(e) Equity Problems

Another problem with marginal social cost pricing would be that it may lead to equity problems. For example, when charging passengers the full social costs of transport this may reinforce problems of social exclusion. Important specific groups that are often considered concern the poor, and the physically handicapped. A broader discussion of equity is given below. As shown above; equity problems may be an unintended side effect of efficiency oriented policies to address transport problems such as congestion and environmental nuisance. A broader perspective is that equity may be the explicit aim of certain transport policies such as the construction of infrastructure in lagging regions. In this case, equity is more than a side effect: it is the main motivation for a certain transport policy. As noted for example by Viegas (2001) and Rietveld (2003) the notion of equity is not unambiguous, however. Equity concepts that arise in the literature and in political debates are:

- ☆ *Horizontal equity.* Comparable individuals should be treated in a comparable way.
- ☆ *Territorial equity.* This results from the notion of individual equity when it is projected on relatively homogeneous regions. For example, comparable regions need to get similar funds for public transport.
- ☆ *Level playing field.* Transport sectors should be treated in similar ways according to taxation, payment for the use of infrastructure, etc.
- ☆ *Vertical equity.* This means that disadvantaged individuals deserve protection. People should be burdened according to their ability to contribute, and this may lead to schemes where taxes are more than proportional with income.
- ☆ *Transport users should pay their way.* As indicated by Gomez Ibanez (1997) this concept is usually interpreted in terms of average costs implying that the collective of all transport users exactly pays for the aggregate costs.
- ☆ *Individuals that are negatively affected by policies need to be compensated.* This principle takes its starting point in the status-quo and says that winners have to compensate losers.

This list of interpretations of the equity notion makes clear that it can be used in various ways by various interest groups. Hence, there is not only a potential conflict between efficiency and equity, but also between various equity interpretations. Consider for example a tax increase in a certain transport sector in order to reach level playing field conditions Such a policy may be opposed by the companies in this particular sector because of the abovementioned status quo arguments.

An important reason why equity considerations are important is that ignoring them may have serious acceptability implications. In democratic societies these implications may have a strong impact on the political feasibility of policies. They are among the main reasons why pricing is a difficult domain in the practice of transport policy making. This multimodal perspective is important for at least three reasons. First, a tendency can be observed that policy makers treat various transport modes in very different ways, that may be explained by historical reasons, but that are difficult to defend on efficiency grounds. This comes close to the grandfathering theme. Examples are the very different treatments of air, water and land transport from the viewpoint of pricing policies. A multimodal perspective helps to avoid biases in transport policies. Second, a multimodal perspective will stimulate learning processes. For example, insights obtained in the domain of airports may be transferable to that of seaports. And third, in line with what is said above on second best pricing, interrelationships between transport modes should be considered.

A second feature of the present collection of papers is the balanced attention that is paid to both equity and efficiency considerations. Efficiency effects are prominent in the contributions of Martin and Betancor, Goeverden *et al.*, and Macharis *et al.*). On the other hand the three road pricing oriented contributions (Allen *et al.*, Ieromonachu *et al.*, and Ubbels and Verhoef) focus on equity and acceptability aspects. An interesting observation is that it is road oriented studies that focus on equity and acceptability. This may reveal the state of affairs in research and policy making. In the road sectorresearchers have for a very long time been exploring the efficiency implications of pricing policies. The major bottleneck nowadays seems to concern equity and acceptability. This is a nice illustration that a balanced treatment of both concerns is needed in research and policy in order to make further progress on this fascinating field of transport pricing. In general, the key findings are:

- ✰ There are substantial disconnect between transport costs and prices in numerous African countries, *e.g.* transport costs in Africa are not abnormally high, but transport prices are high along some corridors, indicating a strong seller's market.
- ✰ Despite many poor efficiency factors (low yearly vehicle utilization rates, aging vehicle fleet, unbalanced trade, etc.), trucking companies in Africa can still charge high price, and have relatively large profit margin along some corridors. Prices are inflated through the whole logistics chain. In a regulated environment, even though profits from freight forwarders may be lower, this activity remains largely profitable. Any overhead or abnormal payment is automatically included in the price for the final user. Moreover, integrated services, which includes shipping lines activities, port operations, logistics platforms operations, freight forwarding and sometimes rail concession usually give a better quality of services but considerably inflate prices on several segments of transport, such as port operations or logistics platforms and these global operators frequently prevent other operators to increase their market share and use their market power to keep prices at very high level. It is explained by the importance of delays on the road on the Northern Corridor (see Teravaninthorn and Raballand regulation is an eminent price determinant hindering the efficiency improvement of the SS Market trucking industry, lacking competitiveness and thus high transport price in Africa.
- ✰ Poor condition of road infrastructure might not necessarily be the most critical factor for high transport costs. The formulation of policy recommendations needs to distinguish between regulated and more mature market environment. In a competitive environment with high traffic volumes, measures to improve road condition and limit fuel prices are likely to yield significant results. In such environment, measures aimed at reducing delays at the border or at weighbridges would also be useful as they would help increase truck utilization. In a regulated environment, like in West and Central Africa, regulatory constraints (formal and informal) must be dismantled because they are the root cause of limited competition, poor service and high transport prices.

9.2. How Acceptable are Transport Pricing Measures?

Transport pricing has been an actively debated topic in the European Union for a number of years. This debate has been stimulated both by the European Commission itself (notably through the publication of the Green Paper on "Fair and Efficient Pricing" and the White Paper on "Fair Charging for Infrastructure Use") and by a multitude of national policy initiatives such as distance-based charging schemes for heavy goods vehicles (*e.g.*, in Switzerland and in Germany), urban and interurban road pricing (*e.g.*, in the Netherlands and the UK) and ecological tax reforms (*e.g.* in Germany). The theoretical aspects of different pricing principles and the economic effects of specific pricing measures have been examined extensively in the scientific literature. However, most debate revolves around the issue of the public and linked to this) political acceptability of pricing measures. Indeed concerns regarding acceptability now constitute the most significant single barrier to implementation, especially in the road transport sector where recent studies suggest that pricing measures are rejected by a majority of car drivers (*e.g.*, Bartley 1995, Jones 1998).

This paper explicitly examines the acceptability of different transport pricing measures and the factors influencing acceptability. It summarises the results of empirical studies carried out in the PATS-project1, a project funded by the European Commission under the 4th framework programme. This project dealt with user charging and taxation measures in the transport sector, covered all modes of transport. The research used both qualitative and quantitative survey techniques. The qualitative surveys took the form of (i) structured interviews with key transport operators and decision makers (key informants) in 9 European countries, (ii) focus groups with the general public, in 3 European countries and (iii) a Delphi survey with key informants, in 5 European countries. These qualitative methods were complemented by an extensive quantitative survey of public attitudes, undertaken in 6 European countries.

The qualitative surveys considered both established pricing instruments such as fuel and vehicle taxation, parking charges, motorway tolling and urban road pricing and more novel instruments such as environmental charges, ecological tax reform and a more differentiated public transport pricing. The results from the initial set of quantitative surveys were used to inform the design of the subsequent public quantitative attitude surveys. The empirical studies presented in this paper have shown that there is an awareness of transport-related problems, in particular of environmental problems and congestion, on the part of citizens, policy-makers and transport providers. Furthermore, we have found that policy-makers and transport providers seem to understand the relations between objectives of pricing, appropriate pricing principles and use of revenues from pricing better than one would have assumed. However, transport pricing is not really understood by citizens, and any increase of taxes and charges is seen as a form of money raising and considered as wrong. The first overall conclusion is that pricing measures attracted in all countries surveyed only rather limited support from citizens. This is not surprising at all:

No consumer loves the idea of being charged (more) for something which was offered for free (or less) before. The empirical studies presented in this paper provided more insight into the factors underlying to this.

- ✰ First, we found evidence of a widespread belief that transport is already too heavily taxed. In the focus groups, for example, both car drivers and public transport users stated that car drivers already pay enough or even too much. This is reinforced by a belief that governments are not necessarily always completely transparent or forthright in their motivations for increasing transport prices. This is often coupled with a feeling of resignation and fatalism that governments will anyway charge transport users without caring for the "victims", exploiting their position as monopolists. In general, people do not feel that they have a voice in the decision process on transport pricing which should clearly be an alerting signal for policy.

- Second, whilst respondents are clear that increases in transport prices will bear heavily on particular groups, such as regular car users, they appear to remain sceptical regarding whether, despite this, the pricing measures will in fact render significant congestion or environmental benefits (due to the likelihood of compensating behavioural adjustments).
- Third, people consider roads and also public transport as basic public services to which people are entitled. An important dimension of this issue is also the existence of reasonable alternatives to the mode/route charged. Both the focus group discussions and the citizen surveys have shown that people value public transport strongly but they consider it as too expensive and/or infrequent to effectively replace private transport.

A second overall conclusion is that while there seems to be a general agreement that the price should reflect the real costs of transport divergent opinions exist on the treatment of 'environmental friendly' modes. For example, some key informants suggested that these modes should pay less while others suggested an equal treatment of all modes. The focus groups argued that pricing was aimed at reducing emissions and insisted that therefore all polluters should pay. The citizens expressed a strong agreement for charging lorries higher than cars and for introducing lower charges for "green modes", however coupled with less support for charging those higher who use roads more than others. The third overall conclusion is that there are significant differences between different countries in terms of attitudes and potential responses to transport pricing measures.

These differences operate in combination with the effects of income and car use to give rise to a very wide range of different views regarding desirable and/or effective pricing interventions. This seems to cast doubt on the viability of a 'one size fits all' approach to the specification of pricing principles and measures and puts a premium on national and local decision making.

A fourth overall conclusion is that well known, traditional pricing instruments were easier to accept than new ones. Variable charges differentiated with respect to time, noise and air pollution, quality of service etc. are preferred. A fifth conclusion is that there is a strong preference for spending the revenues raised from transport user charges in the transport sector. However, there are diverging opinions on cross-subsidisation of other modes of transport. In the citizen surveys a slight preference for using the revenues to improve public transport was stated which might be interpreted as the desire of citizens to have alternative transport modes with a good travel comfort ifprivate car travel is made more expensive. Generally, it seems that the issue of revenue use in other modes than the charged one is highly dependent on the concrete pricing package to be analysed. In terms of acceptability it thus seems sensible to proceed with caution on the issues of cross-subsidisation when implementing pricing measures. Whether and to what degree revenues raised with pricing measures will be spent for other modes should rather be decided a case-to-case basis. The sixth overall conclusion was that, whilst pricing measures were generally unpopular, their unpopularity can be reduced by a suitably configured package of spending measures, focusing, as indicated above, on measures directly related to improving conditions within the transport sector. Conversely, if the revenues are spent in ways that are defuse and render few immediate or linkable benefits; the unpopularity of the pricing measure is likely to be exacerbated. Our studies have also shown that privacy issues are no major obstacle for introducing road pricing.

9.3. Transport Cost, Transport Pricing and Regulation

In approaching the interrelationships among costs, pricing, and regulation, it is assumed that our interest is in allocative efficiency. Relevant questions with respect to rate floors include the cost concepts that appropriately govern private and regulatory pricing policy in competitive situations and the way costs are employed in this crucial aspect of regulation. Since the primary downward pressures on rates are generally conceded to be attributable to the railroads, a special look at their pricing policies is in order. While specific patterns and applications may be changing as a result of new competitive pressures, discrimination remains the hallmark of rate systems. Although frequently inferred, it has certainly not been established through present regulatory tests that inappropriately high price-cost ratios are out of the question. Accordingly, our discussion will be concerned with the appropriate employment of costs by regulators in setting both the upper and lower pricing boundaries within which managerial discretion should be free to operate. It may seem surprising that a matter so thoroughly explored in the literature of economics as the costs which are relevant for pricing should require further attention. But it is apparent from a review of recent literature on transportation economics that the matter is not so straightforward as might be expected.

There is no reason to be concerned about the wild and heated controversy over principles among affected partisan transportation interests. But there is evidence of intellectual disagreement among disinterested students of the transport pricing problem which extends to either or both the governing economic principles and their applicability to transport as a special case. Accordingly, it will be useful to review some recent views on the subject, both to illustrate the areas of disagreement and, hopefully, to rationalize the main points of contention.

The basic issue to which most other questions are subsidiary is whether competitive rates should be "predicated on" or "related to" some measure of incremental or marginal costs as a floor, or whether they should be formally tied to "full costs" through regulatory dicta. The latter contemplates a formal "accounting" assignment to specific services or classes of output of those cost elements which escape measurement under the marginal principle. The most recent statement supporting the unique relevance of incremental costs as a pricing reference is that advanced by a panel of ten economists in the October 1962 issue of the Journal of Business.' Being a member of that group and, along with the others, a subscriber to the principles it advanced, I consider this article a useful starting point for the present discussion. It is essentially a restatement of orthodox principles generally accepted, at least in the abstract, by most economists, and can be summarized very briefly.

- Railroads manifest a greater proportion of unallocable costs than is found in most industries. These unallocable components arise from both the high level of fixed costs and the significant elements of joint and common costs that characterize the industry.
- Of vital importance for the argument and for further discussion is that, while fixity lapses into variability as the time dimension extends and sights are changed from short- to long-run considerations, a substantial proportion of rail costs remain fixed even over time spans far longer than are really relevant for any managerial or regulatory decisions. Since the cost fixities persist and the joint and common components are unaffected by the time dimension, the unallocability of costs is far more than a short-run phenomenon and untraceable elements are extremely important even for relatively long-term considerations. This is far less true of truck and barge line costs.
- The relevant cost reference for pricing decisions is marginal (or incremental) cost since this measure provides the test for the profitability of a pricing decision. In social terms, comparative marginal costs, which provide the economically valid measure of the value of resources used, offer the appropriate guide as to which agency is the most efficient alternative.
- Coverage of the overheads created by the common and joint and the persistently fixed costs can only be achieved through "assignments" to specific services or classes of services that reflect market conditions and thus depend upon service demands. Rates higher than those thus dictated would restrict utilization of the fixed rail plant, artificially inflate shipper costs, and thus contribute to economic inefficiency.
- Derivations of fully allocated costs through arbitrary assignment of the unallocable, as a means for recovering the overheads, is an economically invalid and pernicious device to measure comparative efficiency or to establish relative competitive prices for different modes. But the lack of unanimity on these matters is manifest by some citations from other recent literature. One striking case is found in these-called "Doyle Report" prepared for the Senate Committee on Interstate and Foreign Commerce.2 This study concludes that, with some minor qualifications, rate changes which enhance departures from fully distributed costs are presumptively unreasonably related to costs. Professor George Wilson in his recent volume of essays unequivocably prescribes that competitive rates should be based on the fully allocated costs of the low cost carrier, who would be identified by this measure. Although clearly not accepted as a guide to their own policy recommendations.

Professor Meyer and his associates indicate in their study that one of the historical purposes of minimum rate control was to prevent a carrier (or agency) with lower short-run marginal costs but higher average (or long-run marginal) cost from uneconomically usurping traffic from a competitor. Although other citations might be offered, these bring into focus the significant issues and are frequently cited by others to establish the intellectual respectability of the "full-cost doctrine" in transportation pricing policy. It will be noted that one of the main sources of difficulty is the persistent failure to deal realistically (in terms of relevance for public and private decision making) with the time horizons of cost variability. The Doyle Report gives lip service to the economic attributes and virtues of long-run marginal costs: The recommendation that long-run marginal costs be defined as the appropriate floor to competitive ratemaking is sound long-range policy and will provide a basis for achieving rate structures which reflect the true cost advantages of the diverse modes. But the Doyle Report's definitions of long-run marginal and long-run out-of-pocket costs, which are regarded as virtually synonymous, embrace some puzzling references. At first glance these appear to be simply matters of loose terminology but they colour the use of the concepts and cloud some vital relationships. For example, long-run out-of-pocket cost gives due consideration "to all fixed and indfrectcosts which are incurred because of the extra unit of production." While confusing, this usage might be harmless if it did not influence further thought about cost relations. The vitiation of the important conceptual distinction between fully distributed costs and long-run marginal (or out-of-pocket) costs spawned by the defective definitions begins in earnest with the observation that the fixed costs of the past may or may not be appropriately considered in evaluating "the true fixed costs of additional production." And "if, in fact, they are appropriate guides to fixed costs which are incurred in additional production, it makes little difference in practice whether one refers to fully-distributed costs or long-run marginal or long-run out-of-pocket costs." It is then an easy step to the proposition that "suitable" long-run marginal costs include a return sufficient "to defray all appropriate fixed and indirect costs, thus insuring continued production." no arbitrary allocations will be required. But the "theoretically pure" long run (contemplating complete variability) is, in the railroad case, an abstraction completely devoid of practical significance. As is well-known, large elements of rail plant are not reproduced over very long-time periods, and other investment as well lags far behind output. Substantial elements of rail cost remain fixed over any period relevant for either management or regulatory decisions. The variability that would be required to identify empirically fully-distributed costs with long-run marginal costs would have to be based on calculations stretching out into such an indefinite future as to be meaningless, because of the unpredictability of technological change as well as other factors, such as demand shifts.

The "long-run" concept may be employed in another way which is far more meaningful for railroads–that is, as a time period adequate to permit full accommodation of the plant to any output changes which may be realized. In this meaning, investment levels may change and short-run fixed costs will vary, but it is not necessary to contemplate that all investment costs must have thus varied before a meaningful" long-run" calculation can be made. This is, of course, the conception of the "long-run" that is implicit in the cross-section analyses that undertake to measure cost variability. These analyses cover the wide range of outputs and traffic densities that characterize the nation's railroads and provide empirical support for the reality of "long-term" fixities. Precisely because they do cover such a wide range of densities these analyses tend, if anything, to understate the fixed components and to overstate the long-run marginal costs that are really relevant for economic decisions; accordingly, they represent a highly conservative technique. In their measurements of long-run marginal costs, Meyer and his associates identify costs totalling a "reasonably substantial sum" which "represent unallocable overhead costs in the best empirical meaning of that term." Accordingly, any realistic measure of "long-run" marginal costs will not exhaust all costs to insure against bankruptcy, and uncovered overheads will have to be assigned either arbitrarily or on a demand basis. Furthermore, the untraced ability attributable to common and joint costs will not be washed out regardless of the time horizon of cost measurements. Failure to recognize this empirical distinction between fully distributed and long-run marginal costs is reflected in the final recommendation son cost and pricing of the Doyle Report. One prescription is that "rate changes which result in rates that depart further from fully unreasonably related to costs, unless specific evidence can be presented to demonstrate that such rates are reasonably related to long-run marginal costs." Except for changes in the prices of factor inputs, long-run marginal cost (as realistically defined) will be less than fully distributed costs. Unless a rate is already below marginal costs, any reduction will inevitably bring it more closely into line with that measure. Wilson does not merely flirt with fully distributed cost as does the Doyle Report, but embraces it affectionately. He finds that "the most sensible policy would be to have the rates based upon the fully allocated costs of the low-cost carrier (that is, the carrier whose fully allocated cost is lowest). He is explicitly led to this espousal of fully distributed costs through the traditional rule that "in the long run allcosts are variable and the fixed-cost category disappears." But, unlike the Doyle Report, he does not argue from this to the practical identity of fully allocated and long-run marginal costs. Rather, he realistically recognizes the distinction between train-journeys as the basic output unit and tons or ton-miles as the sales unit; this creates indivisibilities for pricing purposes that persist over the long run. It is these indivisibilities that are arbitrarily prorated in his concept of fully distributed cost. He explicitly rejects what he defines as the usual meaning of fully distributed cost, which involves the allocation of short-run fixed costs, and presumes away any problem of allocating fixities in something greater than the short run.

The earlier comments regarding appropriate and realistic time horizons for the railroad case are applicable here and need not be repeated. However, the pro rata allocation of the indivisible variables requires a closer look. Wilson observes with respect to these indivisible costs that pro rata allocation is both simple and based on precedent attributes that are respectively questionable and irrelevant. These costs appear to be nothing more than members of the common and joint cost family which, as Wilson himself recognizes in another section of his volume, can only be recovered on a no cost (or demand) basis. He is, nevertheless, dubious about the cost allocating function of demand in competitive markets (although the endorsement mentioned is not apparently reserved for non-competitive situations), arguing that there is "a 'critical' point in the operation of an enterprise with large elements of indirect or indivisible costs beyond which further applications of the marginal cost principle can only lead to economic loss; along with this is a corresponding withdrawal of capital or elimination of operations for which rates not much above directly traceable costs have been quoted." It is not particularly helpful, however, to suggest that profit maximizing through demand pricing is acceptable for some portion of output, but that the bulk of sales must be made at prices above the best the market will permit.

This suggestion that competition in transportation is inherently ruinous is of doubtful validity.'7 But even if it is true, there is no particular reason for seizing on fully distributed cost as a pricing reference point, plagued as it is with such serious defects as arbitrariness and circular relationships with volume. If Commission intervention is needed to prevent the ruinous gravitation of rates, other less objectionable standards are available, including, for example, some minimum mark-up above long-run out-of-pocket costs. Despite his formal adherence to fully distributed cost as a competitive pricing guideline, Wilson fortunately does not allow it to intrude too seriously into his substantive policy recommendations. Indeed, he expresses great admiration for "value of service pricing," where the cost and quality of a rival's offerings are the key value measures. Much of this retreat is dictated by wholesome respect for the power of actual or potential private carriage as a market conditioner. It would appear that the fully distributed cost concept, however defined, is as flabby in practice as it is in principle.

Although cited only as an historical argument for minimum rate control and clearly not a reflection of the policy conclusions of the study, a passage in the work of the Meyer group has been loosely used as theoretical support for full-cost pricing. Accordingly, it offers a useful focus for a further consideration of some significant issues of competitive price regulation. This passage refers to regulatory policy which is designed to prevent the movement of traffic by a carrier with lower marginal but higher average costs than a rival. Apparently this policy is concerned with the inefficiency of penalties if a highly transitory cost advantage is converted to higher resource costs.

The railroads characteristically introduce the anomaly of lower marginal but higher average costs than a competitor. But further consideration of comparative cost structures reveals that the marginal cost advantage of rails, predicated on the realistic "long-run" concept that full plant adjustment is made to output increases, is apt to be quite persistent. Marginal cost, it will be recalled, is not a component of total cost at a given output level, but the addition to total costs associated with increasing volume. Social efficiency is, for example, concerned with a comparison of the added rail or truck costs that will result if a given block of output is carried by one mode or the other. Precise determinations of the added costs that might be expected in specific situations undoubtedly require close consideration of the layering and cycling of railroad investment. But appropriately designed cross-section analyses, extrapolating from the cost experience of other companies with higher output levels, provide a valid indicator of the investment requirements and the associated costs that can generally be expected from output increments. Such "long-run" marginal costs provide at least the basic foundation of rail cost measurements from which efficiency may be realistically determined. Cost measures exceeding this level reflect an artificial or contrived degree of variability and thus offer a poor test of comparative efficiency.

As the foregoing suggests, a number of efficiency penalties may be incurred by hewing too closely to the comparative average cost line. Since reproduction of much of the rail investment that is reflected in average costs is long deferred, while truck investment turnover is relatively rapid, it might be worthwhile to take advantage of the free capacity availability reflected in the lower marginal costs of rails. If one or two generations of truck investment (over, say, five to ten years) can be avoided, a clear social saving is realized regardless of longer term outcomes. In addition, increases in output levels may by themselves change the average cost relationships and put the railroad in the low-cost position by that standard as well. Furthermore, technological improvements may intervene to produce the same result. These considerations suggest the folly of excluding railroads from markets where they demonstrate lower long-run marginal costs, regardless of comparative average costs.

Although not so clear cut, the foregoing considerations are applicable in some degree to relatively short-run pricing situations, as well as to the "long-run" in which there is full plant adjustment to expanded output. The relevant pricing horizon for management is clearly determined by the duration of the expected revenue change. According to the panel statement: The decision is governed by such revenue dimensions as the nature and amount of the comtemplated change in volume, the length of the commitment to carry the traffic, the duration and geographic scope of the changed rate, the alterations of the service that might require added investments, and the time period in which changes may legally or practically take place.

Accordingly, profit considerations may dictate a strictly short-run view of costs. This is clearly true if the pricing decision involves a temporary rate (such as one covering the movement of material to a construction project), but extends also to the general case where fuller plant utilization requires a consideration of short-run demand characteristics. There are good reasons to presume that pricing based on short-run marginal costs is not only profitable but is also consistent with efficient resource allocation, particularly since the railroad short run corresponds closely with the long run of a truck competitor. If rail prices predicated on a short-run cost reference are all the market will allow, replacement of the assets required to produce the service involved will be neither socially economic nor privately profitable, dictating substitute investment in an alternative mode. in the meantime, the economy will have enjoyed the use of the available capacity.

However, economies realized from fuller utilization of available facilities must be weighed against the costs associated with such possible short-run dislocations as the elimination of firms with greater long-run efficiency. Accordingly, some "compromise" restrictions on strictly short-run pricing by railroads in competitive situations may be a required function of regulation. As these considerations imply, maximizing returns with rates above marginal costs and below full costs does not occasion a reduction of investment below some socially desirable level. Full cost pricing, which necessarily reflects past investment costs which are unrelated to facility use, does not supply valid investment guidance. Enforcing this pricing policy where such costs are higher than "maximizing" prices (the relevant and interesting regulatory case) will in itself impose artificial restrictions on future investment levels. Sensible management and public policy requires prices that will permit the assets employed to earn as much as the market will allow. Whether the replacement ultimately occurs depends upon the fruitfulness of the maximizing goal, with the decision necessarily made at replacement time. Such a process can be socially objectionable only if the earnings are excessive. The stifling of marginal costs–particularly those associated with reasonably long-term horizons that reflect the investment requirements of traffic increments–cannot serve the public interest.

Despite the established relevance of marginal costs as a measure of comparative efficiency and as a price floor, a system of market-oriented rates predicated on profit maximizing may still be open to question. There may be concern, for example, about unduly high charges in inelastic markets, or excessive discrimination (which is discussed in a later section). But the pricing system contemplated must also pass muster in other important respects, particularly on the validity of the resulting intermodal traffic allocation. The question is whether intermodal price relations may be expected to square with relevant cost relations when prices are designed to maximize contributions above incremental costs. This discussion is independent of the contention that this goal will produce a disorderly gravitation of rates to "ruinous" levels. It may be assumed either that such ruinous pricing is not to be expected or that regulation will prevent it. Where the services of the various agencies in a given market are homogeneous, carrier demand functions are identical and in the maximizing process the carrier with the lower marginal cost presumably can (and will) generate the lower price. As a result, shippers will generally select the low-cost mode to achieve both private and social efficiency. Such service-quality homogeneity is certainly not unknown in some market situations. Cost evidence is more prominent in intermodal than in intermodal cases, appearing in 90 per cent of the former (165 of 185 cases) but in only 71 per cent of the latter (117 of 165).

These percentages varied significantly, however, with the participating agencies. Of the 152truck iritramodal cases, costs were introduced in only 72 per cent. But the ratio rose to 88 per cent in cases involving truck proponents and rail Protestants and jumped further to 95 per cent when the roles of the parties were reversed. It is also noteworthy that motor carriers are less prone to rely on cost evidence than the railroads. In rail rate cases, motor carrier protestants supplied cost evidence in 55 per cent of the cases; but in truck rate cases, the railroads used cost evidence in 80 per cent. Surprisingly, in only 8 of the 252 cases did the cost evidence pertain to fully-distributed costs. Out-of-pocket measures were involved in219 cases, while in the others the report did not specify, possibly because of the vagaries of the presentations. This suggests that the limitations imposed by fully-distributed costs may not be as pervasive as one might be led to believe. While it plays a great part in some important cases, it does not seem to intrude routinely into competitive rate adjustments. For the cases as a whole the denial rate was quite high (70 per cent),with little difference based on whether or not cost evidence was involved.

The direction of the revenue increment is a reflection of the effectiveness of railroad pricing policy in the face of the strategy or any "inherent" market advantages of their competitors. A closer look reveals the strength with which other market forces have actually counteracted increases in market size, as well as the potential traffic-stimulating effect of most rate reductions. In a number of extreme cases, in fact, large increases in market size were so completely counteracted that the volume after the price decrease was less than in the initial period. In the 1959–60 periods, thirty-five rate reductions were accompanied by a market expansion. Despite this favourable market trend, tonnage increases occurred in only fifteen cases, eight of which increased more slowly than the market enlargement. The rate reductions induced an output expansion beyond what would have been expected from increased market size in only seven (or one-fifth) of the cases. But in uneven more extreme instance, twenty of the thirty-five price reductions were associated with an absolute decline in tonnage carried, perversely suggesting a positively sloping demand function and indicating the power of the forces eroding the railroad's market position. The strength of these ant railroad forces can be roughly measured by relating the increased output expectable from the market expansion to the actual tonnage contractions. The sum of the potential revenue increment attributable to market growth, plus the decline actually realized, provides a rough absolute measure of the contributing influences in the twenty cases. Dividing this sum by the first amount reduces this measure to relative terms and produces a "revenue shrinkage ratio" based on the values derived from the factoring processes previously described. A ratio of one denotes a complete offset of the increased market potential, while a higher ratio indicates an even greater revenue depression. A ratio of two, for example, means that the counteracting forces depleted revenues by double the expansion dictated by the enlarged market, despite the reduced prices. The "shrinkage ratios" thus derived for 1959–60 generally range up to nearly seven, with a further jump to seventeen. Extremely powerful market forces are clearly at work here.

9.4. Transport Costs and Discrimination Control

The pricing system obtained by maximizing net revenues in particular markets is, of course, highly discriminatory. When unallocable costs find their way into prices through differentiated demand functions, there will generally be an inequality of P – MC relationships. However, the system does improve utilization of railway plant that is fixed over relatively long time periods, and permits a reasonable degree of managerial discretion in responding to market forces. In this process some prices will be set below average costs at the prevailing output level, creating "downward discrimination." Others will rise above this measure where demand elasticity's permit. In theory, shippers paying below-average and above-average rates benefit along with the transportation companies from these price relationships. The gain to the shippers from the downward discrimination is apparent, while transport companies ostensibly benefit from the greater utilization and lower average costs induced by the relatively low charges. But completely happy results are not assured from the workings of this pricing system. To the extent that transport rates depart differentially from costs, uneconomic patterns of resource use are encouraged in the no transport sectors of the economy. Furthermore, it is not necessarily true that the shippers paying rates reflecting upward discrimination are better off. In other words, in the abstract at least, some form of discrimination control is indicated to limit the unfavourable effect on the non transport sector of the economy and to insure that the rates assessed in the highly inelastic markets do not injure rather than benefit the affected shippers.

In this context, low rates are justified only if they contribute more than higher rates to the support of the system and thus diminish the financial burden on other traffic. The basic social justification for discrimination requires, therefore, that the high rates are less than they would be with non-discriminatory charges. Accordingly, higher than average rates are unjustified if they exceed amounts dictated by the output and unit cost that would be associated with rates represented by uniform relations with marginal costs. To illustrate, assume output(sales) of 50,000 units under discrimination but only 30,000 underuniform pricing, with unit costs of 2.0 and 2.3 cents respectively. In this case, shippers paying more than 2.3 cents would be better off without discrimination; such rates thus assess an unjustified burden of support which violates the basic social rationale of this pricing system. This test contrasts sharply with the present comparative" Since different services are not homogeneous but incur varying direct or out-of pocket costs, discrimination is not eliminated by uniform rates per unit of output, but by a uniform ratio between rates and out-of-pocket costs. In the following discussion, this is the meaning attached to the removal of discrimination and to references to "uniform pricing."

It would be strictly an academic exercise to labor the question of rate ceilings if the present degree of transport competition insures that economic limits of rates are not exceeded. But it is apparent that substitution elasticity varies widely among transport markets, depending on such factors as haul distance and commodity characteristics. Furthermore, the substitutional impact of one major alternative, barge transportation, is rather restricted geographically. It is possible, therefore, that residual monopoly powers (or even regulatory rate floors)may permit degrees of discrimination that exceed socially acceptable standards. Meyer and his associates conclude from their analysis of the demand characteristics of transportation that they "do not set an effective ceiling on the rates of non-competing traffic at a level that precludes monopoly profits."

Although not providing evidence regarding its economic legitimacy, the I.C.C.'s revenue contribution studies describe the sweep of discrimination in rail rates through revenue/out-of-pocket cost ratios for the different territorial movements of the various commodity classes. This behavior will be adopted as a simplifying assumption in this analysis. Based on these cost functions, it is possible to measure the inflation in unit costs from a given percentage decline in traffic and the resulting percentage increase in the average rate level that would be required to meet all costs, including present overhead coverage. For example, if uniform pricing occasioned a 40 per cent traffic reduction, fixed costs would increase by 100 per cent but total unit costs by only 20 per cent. To cover this increase and maintain the same total revenue contribution, rates would have to be set for all traffic at 155 per cent of out-of-pocket costs (where 1960 was 129).

The indicated traffic and revenue effect of uniform pricing at any given ratio of out-of-pocket costs depends upon the price elasticity of rail transport demand in particular markets, and upon the distribution of total traffic volume in terms of revenue-cost ratios. The latter factor is associated with the former since it determines the percentage changes in price (both increases and decreases) which would be involved in moving to uniformity at a given level. For example, output would be less seriously affected by the shift to uniformity if there are substantial blocks of traffic, moving at relatively high rates, which would be benefited by the contemplated reduction (representing the effect of traffic distribution). This would be particularly true if the reductions occasioned sharp increases in the movements of these commodities (representing the elasticity effect). On the other hand, large volumes at lower rate levels, with heavy price increases, would enhance the output reduction, particularly if the associated demands are highly elastic. Traffic distribution is available from the Interstate Commerce Commission's 1 per cent waybill sample, and the rate-cost ratios from the revenue-contribution studies. But since the elasticities are subject to conjecture at this stage of transport market analysis, it is impossible to assess accurately the traffic effects of uniform pricing, or to determine the uniform level of rates which would permit full overhead coverage. Some general indications may be provided, however, by testing the possibility of achieving a traffic level 60 per cent of the present level, with rates uniformly set at 155 per cent of out-of-pocket cost. This is the level that would be required for full overhead coverage, if costs are 70 per cent variable.

On the surface, at least, it appears that the prospects of retaining the required portion of present output with such prices are not at all remote. Although far from conclusive, the foregoing considerations at least suggest that a uniformly established rate ratio of 155 could sustain the level of traffic volume required to support present overhead coverage. But even if it should fall a little short, so that average cost (including overhead) would rise somewhat above the level associated with the155 ratio, these considerations call into serious question the present rates which are at levels substantially exceeding this ratio. This would certainly appear to be true of rates exceeding double out-of-pocket Costs. The foregoing arguments should not be construed as advocating cost based pricing with uniform overhead contributions. The possible effect of this system was considered only in order to visualize its implications for volume, average costs, and appropriate discrimination limits. Discrimination is still a valid railroad pricing device, but its range may quite possibly be excessive in terms of the cost structures associated with present-day traffic densities.

Any rates that exceed a ceiling indicated by the test described here are questionable since the shippers involved are worse and not better off with discrimination. Assuming, however, that the lower rates make their best contribution to overhead, neither public or private purposes would be served by arbitrarily increasing them to some stipulated ratio. Discrimination below the ceiling facilitates overhead coverage and reduces average costs and over-all rate levels, thus truly benefitting those paying above-average rates. But to confine discrimination within the stipulated limits would minimize the distortions associated with uneconomic patterns of resource use. In short, discrimination thus restricted would be both more equitable and more economic. The reduction of excessively high rates may imply a revenue gap and less complete overhead coverage, but this is by no means certain. It is more than possible that many of the offending rates are not only excessive by the standards advanced but are uneconomically high in terms of maximizing overhead coverage. Furthermore, any realistic system of discrimination control should insure that rates at the other end of the axis are not uneconomically low in terms of their revenue contribution potential. The large number of movements failing to cover by dropping the traffic) would enhance overhead coverage by nearly15 per cent. It is likely that much of the traffic involved would, if appropriately priced, make a positive contribution. Furthermore, there must be an interesting volume which does not produce deficits but which is favored by rates which are still uneconomically low. It is possible, however, that further restraints on upward discrimination might restrict over-all earnings and prevent overhead coverage. Such restrictions are certainly not inconsistent with present regulatory policy since the rate structure is honeycombed with maximum rate orders, at least some of which must be of more than token significance. But the desirability of tighter discrimination limits represents a transport policy problem of some magnitude. From one viewpoint, there is no apparent reason for "monopoly" traffic as a class to contribute significantly more to the financial support of the system than does competitive traffic under a system of regulation which is designed to restrict rates to "reasonable" levels. It is precisely in the competitive sector of the economy that, by definition, a "normal" return is earned. In other words, where competition is at work, the market determines financial rewards and investment returns. Fair competition does not generally provide undue restraints on earnings and there is no reason to expect it to do so in transportation, particularly with the usual regulatory limits on the pressures of price competition. If a normal return is appropriately construed as that accorded under competition, there is no reason to have to "make up" anything on that portion of a company's business that is not subject to competitive pressures. If realizable returns are not adequate, in the face of competition, to induce and support additional investment, it should not be made; certainly it should not require the support of monopoly returns in noncompetitive markets.

On the other hand, hewing too closely to the line of more rigorous discrimination control and thus curtailing profit opportunities in the railroad industry may have an adverse effect on innovation and other socially desirable elements of progress. If investments beyond that dictated by competitive market forces (either actual or as simulated by regulation) are regarded as desirable, the available alternatives are, apparently, to subsidize the railroads or permit them to tap markets for all they can get. There are strong arguments for subsidy since the cost of social objectives can thus be born by the economy as a whole rather than by a limited number of consumers and producers or economic sectors. If the subsidy approach is neither feasible nor palatable, and earnings in excess of competitive allowances are indeed required, recourse to highly discriminatory rates is indicated. In this case, the elimination of present maximum rate controls may be called for. If, on the other hand, it is regarded as desirable to limit discrimination in order to strike amore judicious balance between its advantages and its costs, a close look at governing standards and other criteria is required. Although such control has been one of the important functions of regulation, it is questionable whether the historic concepts and methods are adequate to maximize the gains and minimize the costs of discrimination. The comparative method of testing rates was perhaps adequate in earlier years when the forms and procedures of regulation were developed. In the era of lower traffic densities, the legitimate sweep of discrimination was much broader than at present, and the opportunity for developing refined tests was limited by the relative crudity of the. Available analytical techniques and machines. Although perhaps formerly necessary and adequate, the comparative method under which one high rate justifies another is plagued with circularity and provides no standards of any real vitality and validity.

As the need has changed analytical methods and opportunities have improved immeasurably. But rational data exploitation is impossible without a comparable improvement in the concepts of rate control. If maximum rate control is to be continued, the feasibility of employing. But the virtues of the market-oriented prices depend on the quality of the performance of both regulation and private management. Regulation must guard against any excessive discrimination that might arise from the pursuit of maximum profits in separate markets. Effective performance of this function requires far more refined and sophisticated tests–including both concepts and measurements–than have been applied in the past. Costly inefficiencies can also creep into" market-oriented" prices from the private (management) side. Clearly some modifications of the historic monopoly-based "value-of-service" rates are in order. But gratuitous rate reductions that do not improve net revenues (and that are not required to realign rates exceeding the acceptable limits of discrimination) undermine the financial integrity of the transport system; impair investment adequacy, and create inefficient long-run traffic allocations. On the other hand, undue pricing conservatism, particularly on the part of the railroads, may produce the same results.

Unfortunately, the right path is not clearly marked. The zeal for "cost-finding," important as it is, digs up only part of the answer; cost measures tell only the scope of rate reductions that is tolerable, not what is socially efficient or privately desirable. All of the ordinary complexities of oligopolistic (and, with the rate bureaus, duopolistic) price behavior are compounded by the intricate interrelationships of transport costs, rates and total distribution systems. As shippers perceive and measure these interrelationships with greater refinement, distribution requirements become the key to transport demand. It seems to be a safe assertion that despite all of the complexities, far more is known about the measurement of transport costs than of demand. It is questionable, indeed, whether the problem of demand measurement has even been fully conceptualized in the specialized context of transportation since commodity value lost its meaning. The need for greater understanding of demand, and of transport markets generally, is urgent for both pricing and its regulation. This can be accomplished only by broad-scale research efforts in this sector of the economy.

Only through the improved insights thus afforded will there be a reasonable expectation that transport prices–with or without regulation–will produce tolerably efficient allocations. This paper confirms again what has often been remarked about the comparative-statics approach to the theory of cost: the inadequacy of the long-run cost curve of the firm as a tool of economic analysis. Its inadequacies are not in logic but in application. In public utility economics, much more than in analysis of competitive firms, the long-run cost curve repeatedly fails to tell us what we want to know about long-run costs. It almost always turns out that some functional relationship other than the classic one is the one that counts. The classic relationship is that between simple volume, moving in an unchanging grid of time and space, and average cost with all factors fully variable. In railroad economics, the volume measure for such a concept, as Nelson points out elsewhere in this volume, is density of use per mile of railroad. Yet the dimensions of a real problem for decision may include this one only secondarily, if at all, and usually turns on some other–on geographical extensions or cross-connections, secondary systems, bottlenecks, different bases of service, central-station balance in different distributive configurations, or load-factor problems of one kind and another. These produce cost relationships that are hybrids of long- and short-run elements, reflecting several output dimensions and often containing some effects of technological change. These essentially no reversible cost functions, rather than the classic envelope cost curve, are the ones that are relevant to the problems that have actually arisen.

What confronts us now is a problem of transition–from monopoly toward competition, and from value-of-service pricing toward cost-of service pricing in transportation. But, because the railroads still have prodigious economies of scale, indivisibilities, and inflexibilities, they cannot simply be fitted into the new pattern like a small piece into a mosaic. Though their size and scope will be reduced compared with what they were in the days of monopoly, they cannot simply be squeezed down to the appropriate size along the same path they followed when they were expanding. The railroads can be squeezed right out of existence while we are still trying to find what their ultimate role will be in that transportation equilibrium that we have had so much difficulty in visualizing. If the railroads do survive, some indivisibilities must remain, and no doubt some value-of-service prices will remain along with them. Professor Roberts asks the right questions: what will be the size of the railroad system in equilibrium, when all agencies of transport have finally been fitted into the optimum pattern? And will the costs, the rate structure, and volume of traffic appropriate to that equilibrium permit the railroads to survive and prosper? One might suggest additional considerations for the answers. Some kinds of service, and some of the costs associated with them, can be eliminated without affecting the cost-density relationship on trunk lines. Roberts himself notes that dropping the traffic which fails to cover out-of-pocket costs would itself enhance overhead coverage. Some of this traffic would still move at higher rates. Such measures as dropping passenger traffic, or at least handing over commuter services to some welfare agency, would greatly reduce losses without appreciably affecting unit costs on the remaining traffic. The abandonment of spur lines, backwater traffic, and lightly used branches might do likewise without necessarily affecting the density of traffic on main lines, since much traffic is shifting its locus with the increasing geographic polarization of population and economic activity. Where spur lines or weak branch systems are thickly grafted onto a trunk line which would otherwise be strong, some means must be found for lessening the load of debt and capital claims left behind when such unecomonic facilities are abandoned. It seems to me that this would be a most fruitful way to use government financial help, since the only other way to reduce the burden of claims in line with a reduced plant is through bankruptcy and reorganization. Turning now to the question of rate structure, a consensus seems to be emerging among economists who have recently been studying the problem of interagency competition. It is roughly as follows:

I. Short run fully allocated (average) costs are not an appropriate basis of pricing for railroads.

II. The appropriate minimum standard of price for railroads is long-run marginal cost.

III. This is also an appropriate minimum standard for other transport agencies, but because of their particular cost relationships, long-run marginal cost will be very close to short-run. average cost (for all except pipelines–a special case with a decisive advantage in one commodity).

IV. Hence, a rational allocation of freight traffic (excepting pipelines) among competing agencies will be achieved if prices are equal to minimum short-run average costs including a competitive profit for motor trucking and water transport. Railroad rates would depend on the cost-plus-service-advantage of competing agencies for the traffic in question, but the railroad's longrun marginal cost would be a minimum. This much of value-of-service pricing would remain in the equilibrium of transport rates. Of course, the railroads might well be priced out of the market for short hauls, or where speed of delivery is either very important or so unimportant that the waterways' advantages in bulk transport will tell.

But what is this long-run marginal cost that is to be the minimum for railroad rates? Not the same marginal cost that the railroads could look forward to if they were building their entire plant de novo. It is, instead, one of those hybrid concepts. If we take the present plant and shake out the uneconomic parts of it mentioned earlier, the long-run marginal cost is the incremental cost of equipment and long-run maintenance, plus the variable-factor cost, associated with increases in the volume of traffic in the neighborhood of the volume that will move when priced competitively as suggested. Rates equaling or exceeding this "LMC" are economic. It should not be hard to find a reasonably The concept of nontransport distribution cost (NTC) suggested by Roberts is a useful measure of "service advantage." The rule contemplated is that the railroad should be permitted to capture any traffic from other carriers if it's LMC, as defined above, plus its NTC is less than their minimum SAC plus their NTC for the traffic. The rail rate may be brought as low as necessary to accomplish this result, provided it exceeds LMC. Since most types of traffic are diverted from competing carriers only by degrees, this rule for the railroad would amount virtually to value-of-service pricing above LMC, adjusted to the new competitive circumstances which have greatly increased demand elasticity for most railroad transportation services. If this interpretation of the consensus is correct, it also means that the regulatory authority should not use short-run average cost as a minimum standard for rail rates, but it would be justified in preventing cuts below rail LMC down to short-run out-of-pocket cost or below it.

Many critics believe, however, that the antitrust laws, rather than rigid rate regulation, are the best weapon against "predatory" pricing by railroads, and that the regulatory authority often insists on a "fair" division of traffic and preservation of competing vested interests when wholly unjustified by the underlying economic relationships.

What would the suggested principles of rate-making do to the present a location of traffic? What would the equilibrium allocation be, under projected cost conditions? We do not know nearly as much about this as we should, considering that we need to predict whether the railroads could survive under those circumstances and whether massive reallocations and shrinkages in capacity of other agencies of transport might be necessary during the transition. The indivisibilities and economies of scale of the railroad are good reasons for our special solicitude for its traffic volume, to say nothing of its remaining position as the sole available carrier for some types of freight movement. It is much easier to effect a shrinkage in truck or waterway capacity without strong effects on unit costs. If such shrinkage were indicated, ways could be found to ease the transition for those carriers. (It is probably too much to hope in addition for imposition of full user charges and a cessation of uneconomic waterway development).

Many of the estimates of minimum railroad freight volume under flexible pricing that I have seen seem to be on the low side. Some of these estimates are based on inferences from simple commodity characteristics rather than on calculations of cost and transportation demand. Roberts' own calculations are not designed to answer this question, but to show what might happen under strict cost-of-service pricing. We can take this as a minimum estimate or most pessimistic limit of the railroads' share in the total transport volume, assuming uniform rates. If the rules indicated above were followed instead, traffic for the railroads would very likely increase markedly instead of diminishing.

Recent experience suggests that more rational pricing may recapture large volumes of remunerative traffic. One can refer to the multiple-car rates for coal, or the reduced grain rates which have elicited cries of unfair competition from the waterway carriers in the South. Minimum-volume rates and all-commodity rates are also promising, and the lower rates for automobiles combined with technological innovations in automobile carriage have apparently been successful. Rail-truck piggyback service may decisively reduce the railroad handicap of high terminal costs and slow service on package freight and manufactured goods, for all except very short hauls. In spite of Roberts' unfavorable findings on past experience with railroad rate reductions, the general picture of prices and demand elasticities at present leads me to believe that a massive and comprehensive reform of transport price policy would not only increase the rails' share of traffic but also have a strongly favorable effect on their revenues. But I agree with Nelson that these reorganizations, abandonments, rejuvenations and pricing reforms must be put into effect without delay; else the momentum of decline for the railroads may become too strong to reverse, short of government receivership. Complete analysis of transport pricing policy should involve consideration of the implications of pricing arrangements for:

- ✰ Investment in transport facilities,
- ✰ use of existing facilities, and
- ✰ the cost and quality of transportation service.

Although Roberts flirts with cost considerations and occasionally alludes to the relevance of price-cost considerations for investment decisions and industrial location, his paper creates the impression that he is primarily interested in rate-making practices that will encourage fuller utilization of existing railroad plant. He does not seem to be vitally concerned with the implications of pricing from the standpoint of the long-run problem of encouraging efficient investment in transport facilities, nor does he seem to attach much significance to development of transport rate structures which will encourage a more rational pattern of industrial location. The essential difference between Roberts and those–like myself–who advocate the development of rate policies in which rates are encouraged to bear a reasonable relationship to long-run marginal costs is that the disputants are concerned with different aspects of the total problem of transport pricing policy. Roberts deserves credit for emphasizing, as he has done, the danger that emphasis on costs in transportation rates may in practice produce policies that tend to keep rates too high. I agree that policy does better to err somewhat on the side of encouraging flexibility in rate-making. But I regard as either wrongheaded or short-sighted economists who push transport investment problems under the rug and assume that by disregarding the problems of fixed costs in rate-making and letting nature take its course, somehow problems of resource allocation in transportation investment and problems of industrial location will be smoothly resolved. Is it not paradoxical that the competition that has developed in transportation and upon which we economists have come to depend for rationalization of domestic transportation affairs, is infact a competition for railroads from transportation agencies which derive very large amounts of fixed capital from government sources. A crucial issue for modern transport pricing policy and one to which Roberts' analysis does not give sufficient weight is that of investment policy and the relationship which transport pricing bears to investment flows into the transportation system as a Whole.

10

Monetary Policy Impacts on Growth, Inflation and Exchange Rates

10.1. Monetary Policy Impacts on Growth, Inflation and Exchange Rates

To utilize this framework we used data going back as far the 1980s. This is to enable us understand the interaction of key variables, rather than assessing the impacts of changes in monetary policy alone. Here, we were interested in looking at the medium term impact of changes in monetary policy on key variables such as inflation, economic growth and exchange rates. The first VAR model that was used analyzed the following variables treasury bill rates, inflation, exchange rate changes and real economic growth. The second model substituted the rate of growth of the money supply (M2 growth) for the treasury bill rate and retained the other endogenous variables. Variable definitions are presented in Table 10.1

Diskey, using the adjusted fuller tests. In order to check for significance, the endogenous variables were subjected to tests. Pair-wise Granger causality tests was also carried out on all the variables in order to determine which of these relationships are Granger "caused" at standard significant level. The Granger - causality tests was conducted for these relationships. In order to better assess the underlying macro-economic dynamics, we built two simple VAR models and used each of these to describe monetary policy experiments.

Table 10.1: Variable Definitions and Sources

Variables	*Definitions*	*Sources*
Interest Rate	90-Day interest rate: %	Central Bank of Nigeria
Inflation	4-quarter rate of change of consumer price index: %	Central Bank of Nigeria
Growth	4-quarter rate of growth of real GDP: %	Central Bank of Nigeria
Exchange Rate	4-quarter rate of change of exchange rate: %	Central Bank of Nigeria
Price of Oil	Natural log of the $ price of oil	Central Bank of Nigeria
Money Growth	Rate of growth of M2 Supply: %	Central Bank of Nigeria

Source: Adapted from CBN Report, 2014.

10.2. Synthesis of Theoretical Issues and Literature Review Perspective on the Nigerian Financial System and the Nigerian

This section provides insight into Nigerian financial system and how it has aided growth, development, employment and poverty over time. The reforms in the financial system in Nigeria which heightened with the 1986 deregulation, affected the level of financial deepening of the country and the relevance of the financial system to economic growth and development (Nnanna and Dogo, 1998). However, the rapid globalization of the financial markets since then and the increased level of integration of the Nigerian financial system to the global system operation have generated interest on the level of financial deepening that has occurred. The financial system comprises various institutions, instruments and regulators. The According to the Central Bank of Nigeria (CBN) (1993) the financial system refers to the set of rules and regulations and the aggregation of financial arrangements, institutions, agents, that interact with each other and the rest of the world to foster economic growth and development of a nation. According to Nzotta (2004) the financial system serve as a catalyst to economic growth and development through various institutional structures. The system vigorously seek out and attract the reservoir of savings and idle funds and allocate same to entrepreneurs, businesses, households and government for investments projects and other purposes with a view of returns and eventually implicing on poverty and. This forms the basis for economic growth development that is employment generating and poverty reducing. This is largely because the financial system play a key role in the mobilization and allocation of savings for productive, use provides structures for monetary management and the basis for managing liquidity. It also assists in the reduction of risks faced by firms and businesses in their productive processes, improvement of portfolio diversification and the insulation of the economy from the vicissitudes of international economic changes. Additionally, the system provides linkages for the different sectors of the economy and encourages a high level of specialization expertise and economies of scale. Nzotta (2004) further contends that the financial system, additionally, provides the necessary environment for the implementation of various economic policies of the government which is intended to achieve non-inflationary growth, exchange rate stability, balance of payments equilibrium foreign exchange management high levels of employment and poverty reduction. But is this the case in Nigeria?

The Nigerian financial system can be broadly divided into two sub-sectors, the informal and formal sectors. The informal sector has no formalized institutional framework, no formal structure of rates and comprises the local money lenders, thrifts, savings and loans associations and all forms of 'isusu' associations. According to Olofin and Afandigeh (2008) this sector is poorly developed, limited in reach and not integrated into the formal financial system. Its exact size and effect on the economy remain unknown and a matter of speculation. The formal sector, on the other hand, could be clearly distinguished into the money and capital market institutions. The money market is the short-term end of the market and institutions here deal short term instruments and funds. The capital market encompasses the institutions that deal on long-term funds and securities. The regulatory institutions in the financial system are the Federal Ministry of Finance, the Central Bank of Nigeria as the apex institution in the money market, the Securities and Exchange Commission (SEC) as the apex institution in the capital market, Nigerian Deposit Insurance Corporation (NDIC), National Insurance Commission (NAICOM) and the National Pensions Commission (PENCOM). Since 1986, the monetary authorities have adopted various measures aimed at deepening the financial system and reducing the level of financial repression in the system. In terms of flow of funds, the banking system, clearly dominate and has an important impact on the level of economic growth and development. Thus, we can make a distinction between bank based and market-based financial systems. (Stiglitz, 1985, Levine, 2002). These issues have been the focus of theoretical debate for decades. Attempts have also been made to examine whether one type of financial system better explains economic growth development employment generation and poverty reduction in a country than another. Arestis and Luintel (2004) Empirical studies on financial structure and its effects on economic growth have concentrated on the developed economies, especially the United States of America and United Kingdom, which are market based and Germany and Japan essentially bank based. Olofin and Afangideh (2008). Earlier studies include Arestis *et al.* (2001). Hohi *et al.* (1991) and Weinstein and Yafeh, (1998). The Studies points to the fact that the financial structure is important in economic growth. As noted by Olofin and Afangideh, (2008) "the results based on the models of developed countries can only be used as speculation when it comes to economic policy for developing countries. They are not likely to provide a convincing reference point for developing countries, given the differences in their level of development. Moreover critical issues on economic growth development, employment and poverty remain unaddressed. The more developed a financial system and structure, the greater the slice of returns that accrue to financial investors. Financial reforms have been a regular feature of the Nigerian financial system perhaps at responding use of growth and poverty reduction. The reforms have evolved in response to the challenges posed by developments in the system such as systemic crisis, globalization, technological innovation, and financial crisis. The reforms often seek to act proactively to strengthen the system, prevent systemic crisis, strengthen the market mechanism, and ethical standards. Financial reforms in Nigeria dates back to 1952 when the Banking Ordinance was enacted. The deregulation of banking in 1986 provided the impetus for the

Structural Adjustment Programme. The 1986 reform of the financial system saw a policy shift from direct control to a market based financial system, especially as regards monetary management, risk management and asset holding capabilities of the institutions. A number of other reforms followed including the consolidation policy in banking 2005 and insurance 2007. The capital market has also experienced a lot of reforms over the years, especially as regards the capital requirements of the operators, the operational and ethical standards of the institutions and the modalities of the market mechanism.

The reforms in the system impacted positively on the growth of the financial system. But does this growth translate to employment generation and poverty reduction? The system moved from a rudimentary one at inception to a more sophisticated one in 2009 with diverse institutions and operators, diversified financial assets and an enhanced regulatory framework. The reforms have also tried to address the financial gaps in the system, remove rigidities in the system of credit allocation and control and achieve positive real interest rates and greater efficiency by the market operators in the intermediation process. The process of financial sector reform consists of the movement from an initial situation of controlled interest rates, poorly developed money and securities market and under-developed banking system, towards a situation of flexible interest rates, an expanded role for market forces in resource allocation, increased autonomy for The Central Bank and a deepening of the money and capital markets. The link between financial sector stability and growth is, explained by increased market depth, which potentially increases market efficiency. It also reduces risks through the elimination of weak institutions. Financial sector reforms/seeks to develop an efficient framework for monetary management. This encompasses efforts to strengthen operational capacities of the banking system, foster efficiency in the money and securities markets, over-haul the payments system and ensure greater autonomy to the central bank in formulating and implementing macroeconomic policies. Thus, there is the need to deepen the financial sector and reposition it for growth and integration into the global financial system in conformity with international best practices that allows attainment of restored aspirations such as employment generation and poverty reduction. This among other things will require putting in place financial and monetary policy that is employment generating and poverty reducing. Hence, this study. One of the most important policy concerns in most countries is the effect of consolidation of financial institutions on financial system growth and development and the extent to which it will increase employment and reduce poverty.

As the world financial crisis deepens and spreads to the real economy in more countries around the globe, many societies in the developing world Nigeria inclusive that were already experiencing high levels of poverty and underemployment are now threatened to an even greater extent. The World Bank estimated that the increased prices of food and fuel in the first half of 2008 drove 100 million more people into poverty, worldwide. And, now, with the global economic crisis spreading, the situation for poor people in the developing world could become much, much worse. In its January, 2009 IMF predicted that World economic growth in 2009 was only ½ per cent, the lowest level since World War II, and turning negative if measured at market exchange rates.

The ILO Employment Trends Report, issued in January, 2009, warns that: "global unemployment in 2009 could increase over 2007 by a range of 18 million to 30 million workers, and more than 50 million if the situation continues to deteriorate. In the short and medium run, as before the recent crisis, the key will be to generate large scale increases in decent work if developing countries and the world are to avoid a downward spiral into depression into extreme poverty." What is clear is that to design and carry out these employment-oriented macroeconomic policies, the old orthodoxy policy must be abandoned, and policy makers must look for other policy frameworks to inform their macroeconomic and financial policies. Fortunately, these other frameworks and policies do exist, have been used by many successful countries for many years, and can be adapted to the current environment in many counties such as Nigeria. Of course, they must be re-designed and made appropriate to the particular countries and issues involved. Despite the massive challenges facing the unemployed, underemployed and the poor, for the past decade or more, so-called "global best practice" approach to central banking has not focused on economic growth or employment generation; instead, it has pursued formal or informal "inflation-targeting", in which keeping a low rate of inflation – in the low single digits – has been proposed as the dominant and often exclusive target of monetary policy. In this inflation-focused monetary policy, other important goals, such as financial stability, rapid economic growth and employment creation, are seen as inappropriate direct targets of central bank policy; rather they are viewed as hoped for or even presumed – by-products of an inflation focused approach to monetary policy (IMF, 2006). Thus, according to this orthodox approach to monetary policy, the focus of policy is on "stabilization", rather than "growth" or "development", with an implicit assumption that once "stabilization" is achieved, economic growth, employment creation, and poverty reduction will follow. This orthodox view not only specifies the appropriate target of monetary policy, but also the appropriate tools or instruments. The orthodox approach has emphasized indirect, market based instruments of policy, such as short term interest rates, as the primary and often exclusive tool of monetary policy. This is in contrast to the 'direct', quantitative tools often used by central banks which have involved credit allocation methods, interest rate ceilings, and other ways to direct credit to priority economic sectors and goals. In short, the orthodox approach has narrowed both the goals and the tools of monetary policy.

After more than a decade of experience with this inflation-focused, market based approach, the policy record has been rather disappointing, even disastrous for many countries. Perhaps Nigeria inclusive (Ball and Sheridan,2003). While it might seem obvious that stabilization focused central bank policy represents the only proper role for central banks, in looking at history casts serious doubt on this claim. Far from being the historical norm, this focus by central banks on stabilization to the exclusion of development represents a sharp break from historical practice, not just in the developing world but also in the now developed countries as well (Epstein, 2007a). In many of the successful currently developed countries, as well as in many developing countries in the post-Second World War period, development was seen as a crucial part of the central bank's tasks. Now, by contrast, development has dropped off the "to do" list of central banks in most developing countries As the world digs itself out of this economic crises and establishes a new macroeconomic framework, there should be a return to the historical norm of central bank policy: in particular, employment creation, financial stability and more rapid economic growth. This study will provide the pathway to this important economic return in Nigeria. The Central Banks need not, and indeed, cannot be the only institution having an employment generation role. Fiscal policy, industrial policy, labor policy and others must play a crucial role as well. But, in most developing countries, central banks need to cooperate with other institutions by doing much more than simply keeping inflation rates in the low single digits. To bring this about, many institutions will have to get involved.

10.3. A Critique of Inlation Targeting

According to its advocates, "full fledged" inflation targeting consists of five components: absence of other nominal anchors, such as exchange rates or nominal GDP; an institutional commitment to price stability; absence of fiscal dominance; policy (instrument) independence; and policy transparency and accountability. (Mishkin and Schmidt-Hebbel, 2001, p. 3; Bernanke, *et al.*, 1999). In practice, while few central banks reach the "ideal" of being "full fledged" inflation targeters, many others still focus on fighting inflation to the virtual exclusion of other goals. The overriding announced goal of inflation targeting central banks is typically "price stability", usually defined to be an inflation rate in the low single digits. In addition, inflation targeting is usually associated with changes in the laws that enhance the independence of the central bank (Mishkin and Schmidt-Hebbel, 2001, p. 8). As we see below, this relatively rigid commitment to a very low inflation rate can lead to monetary policy that is too restrictive, leading to excessively high real interest rates, sub-optimal growth of GDP and employment, and an over-valued exchange rate. The major claim made by advocates of inflation targeting is that it will:

- ☆ Reduce the rate of inflation
- ☆ Enhance the credibility of monetary policy
- ☆ Reduce the *sacrifice* ratio associated with contractionary monetary policy
- ☆ Help attract foreign investment and encourage domestic investment

The evidence on these claims is mainly in the negative. It is true that countries that adopt inflation targeting often achieve lower inflation rates. But there is strong evidence that this decline in inflation might not be due to inflation targeting itself, but rather to the general declines in world-wide inflation pressures, (due for example, to major increases of the global labor and increases in global production capacity due to extraordinary export oriented growth in China and the fall of the Berlin Wall) or to a simple reversion to a more normal inflation rate (Ball and Sheridan, 2003; for a contrary view, though, see IMF 2006). In addition, most evidence indicates that inflation targeting central banks do not reduce inflation at any lower cost than other countries' central banks in terms of forgone output. That is, inflation targeting does not appear to increase the credibility of central bank policy and therefore, does not appear to reduce the sacrifice ratio. (See Bernanke, *et al*, 1999 and Epstein, 2000, for detailed surveys of the literature).

Typically, central banks that reduce inflation do so the old-fashioned way: by raising interest rates, causing recessions or slow growth, and contributing to job loss. Moreover, there is no evidence that countries adopting inflation targeting manage to attract more foreign investment so Why the Focus on Inflation? There is a further, basic problem with inflation targeting and the neo-liberal approach to central bank policy more generally. Why is there such a focus on fighting inflation to the exclusion of other goals? And even if fighting inflation is important, as we agree it is, why pick a target in the low single digits? As reported in Bruno and Easterly (1996) and Epstein (2000) there is a great deal of evidence that moderate rates of inflation, inflation up to 20% or more, has no predictable negative consequences on the real economy: it is not associated with slower growth, reduced investment, less foreign direct investment, or any other important real variable that one can find. (Also see Pollin and Zhu, 2006). Some types of price increases, usually associated with "supply shock" inflation – especially those involving price increases for basic necessities, can have very serious negative consequences for the poor. But in this case, it is not clear that restrictive monetary policy is the best response. Acquiring more supply of the commodity "in shock" would generally be a preferred policy response.

Apart from growth effects, however, many economists have argued that inflation harms the poor more than the rich. Hence, on distribution grounds, inflation reduction into the low single digits should be a priority. While more research is necessary to fully investigate this claim, important recent work by Jayadev calls this conventional wisdom into question (Jayadev 2006, 2008) by showing that in a sample of richer and middle income developing countries, poorer and working class people prefer the government reduce unemployment than fight inflation. In addition to these concerns, Braunstein and Heintz (2008) find that disinflationary monetary policy in developing countries such as Nigeria, like that undertaken in connection with inflation focused monetary policy regimes, has a disproportionately negative employment effect on women, relative to men. Braunstein and Heintz conjecture that this might be due either to the sectors while tight monetary policy harm most directly. For example perhaps export sectors employ a higher ratio of women, and they are hurt disproportionately by tight monetary policy which raises interest rates, attracts foreign capital and leads to an overvaluation of the real exchange rate. Another possibility is that, because of discrimination, women are "first fired" in a recession. Both of these are just conjectures, however, and this empirical finding calls out for further explanatory research. Still, Braunstein and Heintz's excellent work is highly suggestive that there may be important gender effects of monetary policy, suggesting that this area needs much more research than it is currently receiving. Of course, all agree that very high levels of inflation, above 40%, can have every serious impacts on growth and possibly the distribution of income. Even inflation above 20% can create problems, according to recent data. But there appears to be very little justification for monetary policy oriented toward keeping inflation in the low single digits, especially when employment and poverty are significant problems as is the case in Nigeria. Still, it is important to recognize that some types of inflation unambiguously hurts the poor. Cost-push or supply side inflation of basic necessities, such as food, oil, and other basic supplies, can have devastating impacts on the poor. This type of inflation, in fact, is often the dominant type of inflation in most developing countries. However, it is unlikely that tight monetary policy, in and of itself, is the best response to this type of inflation.

10.4. The Performance of Liberalized Financial Systems in Developing Countries

The dominant financial policy in recent years in most developing countries, has focused a financial liberalization. But while financial liberalization has had some successes, for example, financial development in Costa Rica and some other countries, it has also had numerous failures. On the positive side of the balance sheet, liberalization has furthered the integration of developing countries into global markets. This has meant that some large firms, especially in the context of privatization programs, have received significant finance through the internationally-integrated capital markets created or expanded following financial liberalization. The finance provided to these firms has often been cheaper than that available via bank loans.Still, there is little evidence that the growth of capital markets increases access to or lowers the cost of finance for those entrepreneurs that have long confronted severe capital constraints. Indeed, as McKinley shows, commercial banks have concentrated their activities in major urban areas of developing countries. He goes onto explain that "although aggregate statistics of financial deepening might have improved following financial liberalization, access to credit has become, if anything, more unequal. The rural population remains deprived of credit in most countries, and is likely worse off compared to the access to credit that state-owned agricultural banks had previously provided". McKinley (2005) concludes that, in the African case, the private sector has even less access to credit after financial liberalization than before.

Large, foreign-owned banks come to play a greater role in the domestic financial system following the removal of restrictions on their presence. Large, foreign-owned banks that generally enter developing countries after liberalization are often not responsive to the needs of small- and medium-sized enterprises SMEs. There is a large body of empirical evidence demonstrating that domestic financial liberalization has failed to deliver many of the rewards claimed by its proponents. Financial liberalization has created the climate, opportunity and incentives for investment in speculative activities and a focus on short-term financial gains. In a number of cases, this has "crowded out" long-term developmental returns. Granted, the creation of a speculative bubble may temporarily result in an increase in investment and overall economic activity. But an unsustainable and financially fragile environment or what Grabel (1995) terms "speculation-led development" is unlikely to be in the long-term interest of developing countries. One channel by which the speculation-led development induced by financial liberalization worsens the situation of the poor is through its effect on financial fragility, and ultimately on the prevalence of currency, banking and overall financial crises. Many empirical studies find that financial liberalization, including the liberalization of capital accounts, often leads to currency and banking crises (see Grabel, 2003a,b,c and Reinhart and Rogoff, 2008a and 2008b). Financial liberalization also can worsen the situation of the poor by increasing income and wealth inequality and by aggravating existing disparities in political and economic power. This is because only a very small proportion of the population is situated to exploit the opportunities for speculative gain available in a liberalized financial environment. Speculation-led development often creates a small class of financiers who maintain greater ties to financial markets abroad than to those in their own country, and it is also associated with a shift in political and economic power from non-financial to financial actors (Grabel, 2002).

11
Challenges with Current Financial Structures

11.1. Challenges with Current Financial Structures

There are many challenges because the financial sector in many developing countries are currently facing challenges that imparts need to be explained to put in place employment generating and poverty reducing monetary framework Despite financial liberalization, and in some cases even because of it, financial structures in many developing countries are not fostering development, employment generation and poverty reduction. They also face the following challenges listed below:

- *Real Interest Rates and Interest Rate Spreads are high:* Real interest rate spreads remain very high in many developing countries, and especially in sub-Saharan Africa. The banking systems are highly concentrated, hold high levels of excess reserves and lend primarily to very large businesses and wealthier consumers. (See Heintz, 2008). Nevertheless, there are examples of well performing financial sectors in some developing countries. Most of these retain a significant degree of regulation (Stallings and Studart, 2006).
- *Credit Creation is too low:* In many poor countries, high spreads and real interest rates are associated with relatively low rates of credit creation.
- *Credit and Capital Flows are: Pro-Cyclical*: When capital flows do come into many developing countries, they are often short-term and pro-cyclical, and sometimes associated with "sudden-stops". These can, in turn, lead to financial crises that can have devastating economic impacts, often especially for women and the poor (Ocampo, 2003).

- *Insufficient Amount of Long-Term, Patient Capital:* The short term and unstable flows of international capital represent a more general, severe problem facing many developing countries – namely, there is a dearth of long-term, patient capital to support long-term investment (Stallings and Studart, 2006). Much capital is of a highly short term nature. As a result, long-term productive investment is extremely difficult and costly to finance.
- *Insufficient Capital for Small and Medium Enterprises and the Poor:* Finally, there is a lack of capital for small and medium enterprises and the poor in most regions of the world Nigeria inclusive (Stallings and Studart, 2006). This problem flows from many of the "stylized facts" described earlier: the high real interest rates and interest rate spreads, the mis-allocation of global financial resources, the short-term and pro-cyclical nature of international capital flows, and the absence of long-term, patient capital. An additional reason is the ability of banks to allocate their assets to high yielding, relatively safe, government securities, which may crowd out lending to SME's and others.

  In many countries, the preferred solution to this problem has been a heavy dose of financial liberalization (Nissanke and Aryeetey,1998; Honohan and Beck, 2007). Recently, liberalization has been combined with the attempt to create a regulatory framework which would monitor and guard against financial instability, while creating space for private financial investment and enterprise, often from abroad. But as the current crisis demonstrates, financial liberalization by itself will not be sufficient to do the job. (Serieux, 2008; Honohan and Beck, 2007; Heintz and Pollin, 2008).
- *Insufficient attention to the problems of rural borrowers:* Even though people living in rural areas make up the vast majority of the world's (and Nigeria's) poor, they are the most underserved in terms of financial services solutions presented below should be designed specifically with the rural poor in mind.

11.1.1. Central Bank Support for Developing Banking and Small-Scale Financial Institutions

The structure of Central Bank support for development Banking and small scale financial institutions may be crucial for a nation's climate. Central Banks have historically played a much larger role in supporting longer-term economic development goals than has become fashionable more recently Epstein (2007a). This has been true both in developing countries and in the now developed countries. Central banks, acting as agents of development, have bought long-term bonds for infrastructure development; kept and administered differential interest rate ceilings in support of favored sectors, such as housing in many developed countries (Epstein, 2007a); used their discount window to discount loans for specialized lending institutions, such as microcredit institutions provided efficient or even subsidized financial services for specialized institutions playing important social roles. These policies have not always been successful, but if planned carefully with monitoring and safeguards, have an important part to play in mobilizing and re-deploying finance for development purposes. For example, central bank support for the Grameen Bank in Bangladesh and for rural development banks in Indonesia have been very successful (Asian Development Bank, 2000). By supporting development banking and small-scale financial institutions, Central Bank will be promoting development that is employment generating and poverty reducing.

While moderate levels of inflation appear to have no significant costs, using central bank policy to fight inflation does have costs, and some evidence suggests that these costs are felt more by some groups than others. In particular, as shown by Braunstein and Heintz (2006), dis-inflationary monetary policy in developing countries, like that undertaken in connection with inflation-focused monetary policy regimes, has a disproportionately negative employment effect on women, relative to men. That there may be important gender effects of monetary policy, suggesting that this issue needs much more research than it is currently receiving although this is not the main issue in this study. Their empirical exercise explores the effects of inflation reduction on women's and men's formal employment. They compiled data for 51 "inflation reduction episodes" in 17 low- and middle-income countries. To assess the employment effects of inflation reduction periods, they looked at actual employment trends during each inflation reduction episode, disaggregated by gender, and compared these to long-run employment trends. They also examine the indicators that suggest how monetary policy responded during inflation reduction episodes using a similar approach. In looking at the impacts of disinflation on employment, they distinguished periods when both inflation rates and employment fell, relative to long run trends, from those periods when inflation rates fell but employment expanded. They use the term "contractionary inflation reduction episode" to refer to the inflation reduction periods when employment contracted and "expansionary inflation reduction episode" to those when employment expanded. According to this study:

- ☆ If employment contracts during an inflation reduction episode, it is likely that women will experience a larger loss of employment, in percentage terms, than men. In the majority of cases, contractionary inflation reduction has a disproportionately negative impact on women. However, during inflation reduction episodes in which employment expands, the gender-specific impact is ambiguous.
- ☆ When central banks respond to inflationary pressures by raising real interest rates above the long-run trend they are more likely to experience a slow-down in the growth of employment relative to those countries that keep interest rates in line with or below the long-run trend, with concomitantly higher losses for relative female employment.
- ☆ They find that real exchange rates (RERs) seem to impact the gender bias of contractionary inflation reduction episodes. In all cases where women experienced relative employment gains during employment contractions, exchange rates either depreciated or showed no deviation relative to long-run trends.

- ✰ Tightening the real money supply also seems to be negatively associated with employment in general and women's employment in particular.

These results suggest that contractionary monetary policy aimed at reducing inflation often has a disproportionately negative impact on women's employment, an effect that may be eased by maintaining a competitive exchange rate. Moreover, inflation reduction episodes where real exchange rates did not appreciate were less unfavorable to women's employment. One of the major socio-economic issues confronting Nigeria is the high and rising number of the unemployed in the country. The problem is even more acute among the youths. The unemployment rate in Nigeria was single digit and on the average less than 4 per cent during the period 1990-1998 (Figure 11.1). The rate jumped from a single digit of 3.5 per cent in 1998 to 17.5 per cent in 1999 (a five hundred per cent increase). Since this period, the unemployment rate has remained double digit. The salient facts about youth unemployment in Nigeria can be summarised as follows:

(a) Out of the total population of 169.15 million in 2012, the unemployed were 18.9 million, of which 11.1 million (58.8 per cent) were youths of between the age of 15 and 34 years;

(b) About 38.7 per cent of the 11.1 million were within the 15 to 24 years age bracket, compared with 20.1 per cent between the age of 25 and 34 years;

(c) 56.8 per cent of the unemployed youths aged 15 and 34 years have less than secondary school education, they are less skilled and inexperienced.

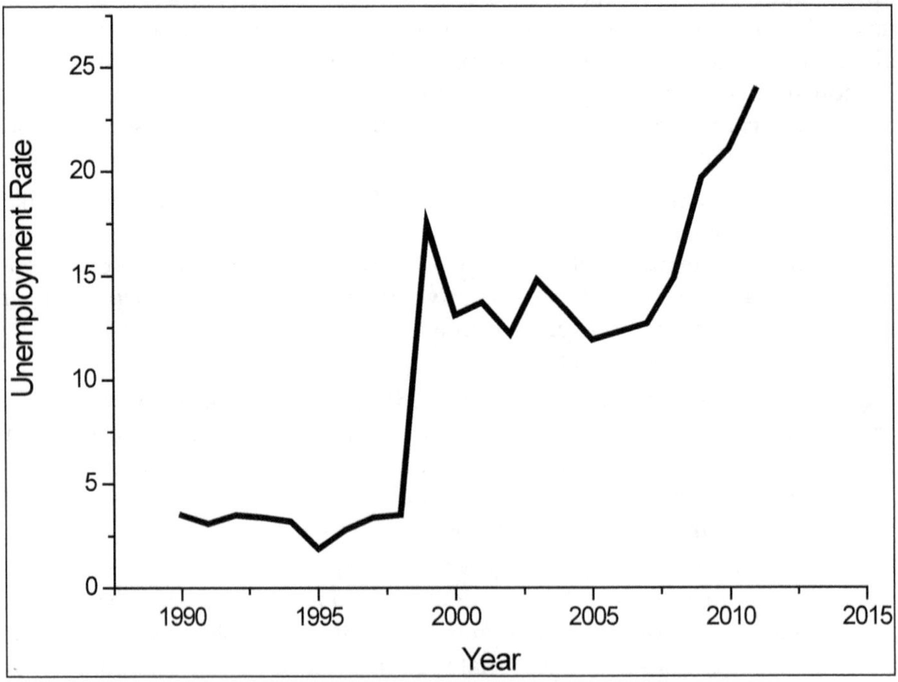

Figure 11. 1: Unemployment Rate in Nigeria, 1990-2011.

***Source*: Generated from National Bureau of Statistics database.**

Nigeria has a diverse social security system which does not sufficiently address the needs of the poor and the unemployed. The government has therefore introduced several intervention programmes to address the problem of youth unemployment. The programmes include the National Poverty Eradication Programme (NAPEP), the National Directorate of Employment (NDE) Programme, Youth Employment Programme (YEP), job creation programmes in Niger Delta Development Commission (NDDC), Millennium Development Goals (MDGs), and the Subsidy Reinvestment and Empowerment Programme (SURE-P). Although these programmes target critical needs, they fail to fully address the needs of the 11.1 million unemployed youths in 2012.

On the basis of empirical analysis and literature reviews undertaken in this study, the conclusion and recommendation made are discussed below. Some Financial Solutions:

- ✰ On important approach, then, is to combine the resources available to formal sector financial enterprises and government financial institutions such as the central bank with the access, innovation, and lending experiences of microcredit institutions (Heintz, 2008; Atieno, 2001; Aryeetey, 2001; Pollin, Githinji, and Heintz, 2008). There are several ways to do this that have been suggested in the literature, and also put into practice, if only on a small-scale basis. Among the most promising are those that involve programs in which formal sector institutions such as commercial banks or central banks take positions in cooperative, microcredit and other development-oriented financial institutions, either by lending to them and or by taking equity positions in them. To be successful, these programs usually involve government regulations and subsidies that generate carrots and sticks for formal sector institutions to extend credit to these smaller institutions that then on-lend to poor and/or small borrowers with good potential.
- ✰ Another way to raise lending and investment to scale so that it can have a transformative impact is to create *development banking institutions* that take direct positions in final borrowers who are engaged in activities that will generate large numbers of good jobs and high value- added, and contribute to the structural transformation of the economy.
- ✰ Mobilizing financial resources and allocating them to productive units in key sectors could be a crucial component of the development strategy for developing countries such as Nigeria, suggested are financial policies and innovations that could be implemented, such as asset-backed reserve requirements, development banking, and loan guarantees, that can help generate more investment in key sectors to increase employment and incomes for the poor (Pollin, Epstein, Heintz, and Ndikumana, 2006; Epstein and Heintz, 2006; Pollin, Githinji, and Heintz, 2008; Epstein and Grabel, 2006).

Several policies along these lines that can be used to help re-deploy and transform the financial sector to facilitate creation of more good jobs, raise productivity, and reduce poverty are three:

1) Asset backed reserve requirement (ABRR); 2) loan guarantees; and 3) development banking.

Formal Sector Banks

Asset-backed reserve requirements (ABRR)

As discussed in more detail in Pollin *et al.* (2006, 2008), with a system of asset-backed reserve requirements (ABRR), banks are required to hold reserves against their income earning assets. However, they are allowed to hold smaller required reserves on assets that are identified as contributing to employment generation and poverty reduction. This would encourage banks to lend directly to high-priority sectors. Similarly, securities issued by smaller institutions that are experienced in making loans to small/poor borrowers or small and medium sized enterprises, microfinance institutions (MFIs), and rural banking cooperatives can have preferred reserve requirements as well.

Loan Guarantees – Loan Guarantee Scheme

This section draws heavily on Epstein and Heintz (2006); Epstein (2007); Heintz and Pollin (2008); Pollin *et al.* (2006, 2008). Pollin *et al.* (2008), and Atieno (2001), who discusses such a program for Kenya. A loan guarantee program could help to mobilize resources from commercial banks for on-lending to microcredit and other institutions who have the knowledge and experience to on-lend to cooperatives, small entrepreneurs, and businesses for productive activity, or to make direct loans to the final borrowers, if the commercial banks so choose. The basic idea is that the government would guarantee a certain percentage of the loan, thereby reducing the default risk faced by the bank; this can help to substitute for collateral which most borrowers in the target group will not be able to provide. This would also help keep the interest rate charged to the final borrowers lower.

There has been extensive experience with loan guarantee schemes spanning many countries over many years. Green (2003) estimates that there have been over 2000 loan guarantee programs of various kinds in almost 100 countries, including most OECD countries as well as in many developing countries. Evidence on their success is mixed, as one would expect with some many schemes in so many places. (Green, 2003; Beck, *et al.*, 2008; Honohan, 2008). In the United States and Canada, major loan guarantee schemes for small business are seen as quite successful. (Green, 2003). Many programs established in Latin America in the 1980's and in some parts of Africa then a bit later, however, are commonly seen as having failed. (Green, p. 23.) Still, according to a UNIDO sponsored study, the best developed African schemes are in Egypt, Morocco and South Africa. (Green, 2003). While there are many types of loan guarantee programs, we discuss one such type here (see Honohan, 2008, Beck., *et al.*, 2008 and Green 2003 for a discussion of other types of programs.) Our suggested program works as follows. The government chooses to guarantee a certain portion of loans from commercial banks or other lenders to borrowers who will achieve program goals, such as generate more decent jobs, increase productivity, and improve the quality of jobs, or invest in innovations or infrastructure to reduce the cost of inputs or consumption goods consumed by the poor, and thereby have strong multiplier effects through the economy to raise the standard of living of the poor. The guarantees may also underwrite a program by commercial banks to lend to microcredit institutions who will then lend to final borrowers.

In setting up such a program, there are several issues that must be addressed:

- ✰ How to determine to whom to lend
- ✰ What should be the rate of loan guarantee?
- ✰ How to monitor the program to avoid corruption, excessive moral hazard, and inefficiency

How to Determine to Whom to Lend

Indirect lending to development institutions. The key here is to start with development finance institutions that have the knowledge, track record, and experience to identify final borrowers that can use the credit productively to create decent jobs, directly or indirectly, for themselves and, on a larger scale, for others in their community. Careful attention needs to be paid to the organizational structure of the program, and it may vary from region to region.

Direct lending by commercial banks. If the large banks want to start lending directly to such borrowers, then, since they do not have a track record of doing so, it would work best if they were to develop a business plan and a "Decent Employment Impact Statement" that makes clear how many jobs, and at what wages and benefits, would be generated by their loans, and, if possible, some clear idea of the domestic linkages and multiplier effects.

What should be the rate of loan guarantee?: In choosing the rate of loan guarantee, there are a number of considerations that must be taken into account, including moral hazard, and the cost of the program to the government. The problem of moral hazard dictates that the government guarantee must be less than 100% of the loan, and possibly significantly less. To reduce moral hazard (and the cost to the government), other mechanisms that substitute for collateral should also be developed,

How much would this program cost the government?: Setting aside administrative costs, the major cost to which the government would be exposed is the cost of defaults. How large a cost would this be to the government? This depends on the size of the program and the default rate.

What changes could reduce the cost? There are several:

1. Reduce the rate of default: If the rate of default was lower, then the cost to the government would be lower for any rate of guarantee. Having a good system of monitoring, and good programs of selecting and working with borrowers, is crucial to reducing the default rate.
2. Reduce the base rate. There are several ways of reducing the base rate:
   - ✰ Less tight monetary policy can lower the base rate, for any given inflation rate.

- ☆ More competition in the banking system can lower the rate, which is quite high by international standards.
- ☆ More competitive and better organized auctions of government debt can lower the base rate.

Monitoring and anti-corruption protection: As discussed in detail in Pollin, Githinji, and Heintz (2008), a potential problem with any program is corruption. In business as well as government, monitoring and accountability issues will be crucial. In the case of the loan guarantee program, it is quite easy to see the form that corruption could take, especially where the government guarantees a large percentage of the loan: the lender and borrower could arrange a false loan; the borrower could default, receive the payment from the government, and split the proceeds. Of course, close monitoring and accounting could, in principle, prevent such corruption from occurring. But monitoring itself has costs and there are advantages to add complimentary measures to monitoring schemes to reduce the incentives for corruption. Pollin *et al.* (2006, 2008) have designed an anti-corruption system based on escrow accounts and "whistleblower" awards that can help cut down on the likelihood of significant corruption. These are described briefly in the following paragraph.

Incentive-based Monitoring

Escrow accounts. To reduce the incentive for corruption, the government could implement a system of escrow accounts. The account would work as follows: the lender places a percentage of the loan in an escrow account with the government or central bank. If there is a default, the lender will receive the amount placed in escrow only after a thorough investigation of the case to establish there was no corruption involved. If there is no default, then the borrower receives the amount at the end of the period, which can then be rolled over as part of the new loan with minimal administrative interventions (forms, filings, etc.) or can be repaid to the lender. Such escrow accounts clearly cut down on incentives for corruption, but this is at some cost in terms of the availability of funds to the borrower. *Histle blower incentives.* This is another incentive-based monitoring measure suggested by Pollin *et al.* (2007), based on an idea promoted by Dean Baker. His idea is to create large penalties for fraud, combined with high rewards for "whistleblowers"

How to pay for the loan guarantee programme: Loan guarantees can be attractive, at least on a start-up basis, to foreign aid donors. But donors aside, a key source of funds could be a dedicated tax which underwrites this set of financial incentives for economic transformation. Of course there will be many worthy demands on tax revenues. But the demands that should be considered, is a loan guarantee programme to help generate investments in good jobs for the long term.

Development banking initiatives: Development banking initiatives were crucial in the successful development of industrial economies of Asia and Latin America, what Alice Amsden calls "The Rise of the Rest" (Amsden, 2001), but the reputation of "development banking" experiences in Africa has been seriously tarnished. Yet the major challenges for development and financial policies – and the insufficiency of financial liberalization as a solution - call for a reconsideration of numerous policy options, including the revival of development banking (Ndikumana, 2007). Success of development banking initiations edit in Africa depends on: integration of development finance institutions into the national development agenda; appropriate governance; efficient operating procedures; minimizing non-interest barriers to access to credit; and innovative strategies for the mobilization of stable long-term funds (Ndikumana, 2007).

Alternatives to Inflation Targeting

This is required especially in countries where unemployment and poverty are on the increase such as Nigeria. There are alternatives to inflation targeting that hold out the promise of a more employment friendly monetary policy and that can be tailored to the particular circumstances and needs of different countries. These were developed by a team of researchers working on a PERI/Bilkent project on alternatives to inflation targeting, as well as a United Nations Development Project (UNDP) sponsored study of employment targeting economic policy for South Africa. The countries covered in PERI/Bilkent project are Argentina, Brazil, Mexico, India, The Philippines, South Africa, Turkey, and Viet Nam. It is a case of one size does NOT fit all. A range of alternatives options are modest changes in the inflation targeting framework to allow for more focus on exchange rates and a change in the index of inflation used, to a much broader change in the overall mandate of the central bank towards employment targeting, rather than inflation targeting. Some of the alternative policies focus exclusively on changes in central bank policy, while for other countries, changes in the broad policy framework and in the interactions of monetary, financial and fiscal policy are proposed. Some incorporate explicit goals and targets, while others prefer more flexibility and somewhat less transparency. But all of the studies agreed that the responsibilities of central banks, particularly in developing countries, while including maintaining a moderate rate of inflation, must be broader than that, and should include other crucial "real" variables that have a direct impact on employment, poverty and economic growth, such as the real exchange rate, employment, or investment. They also agree that in many cases, central banks must broaden their available policy tools to allow them to reach multiple goals, including, if necessary, the implementation of capital management techniques (Ocampo, 2002; Epstein, Grabel and Jomo, 2006).

Real Targeting

The employment targeting approach is an example of a *real targeting* framework. A real targeting framework for monetary policy adds one or more important real variables, such as real GDP growth, or a stable and competitive real exchange rate (SCRER), or "full employment", to nominal variables, such as the rate of inflation, as a goal of monetary policy. The so-called Humphrey-Hawkins law in the United States, which commits the U.S. Federal Reserve (FED) to maintaining both "price stability" and "high levels of employment" is an example of a *real targeting* framework, though, admittedly a fairly weak one in that it does not specify exactly the meaning of "price stability" or "high employment".

A *real targeting* framework has a number of important advantages.

- First and foremost, it places front and centre the economic variables that have the most immediate and clearest association with social welfare. The Central Bank of Nigeria (CBN) must identify this target and then communicate with the public about the actions it is taking to reach it.
- The Central Bank may develop new tools of monetary policy. For example, if a central bank is tasked with hitting an employment target subject to an inflation and balance of payments constraint, then – in addition to interest rate policy - it might explore asset allocation strategies to encourage banks to lend more to high employment generating uses, and capital control techniques to manage balance of payments problems.
- A real targeting approach lends itself naturally to a more democratic, transparent and accountable central bank policy that serves the genuine needs of the majority of countries' citizens, rather than the minority that typically benefits from the combination of slower growth, low inflation, and high real interest rates.
- Another is to tailor monetary policy to the specific needs of different countries. For example, if a country has a particular problem with generating good jobs or more jobs in a particular region of the country, or the entire country as it is in Nigeria, then the real targeting approach can accommodate such needs. Of course, the central bank will find it difficult to use "macroeconomic" tools to target particular groups or particular regions, but the central bank will have an incentive and permission to work with other ministries and the private sector to develop tools and mechanisms to improve employment opportunities for these groups and regions.

One important objection to a "real targeting" approach to monetary policy is the claim that monetary policy does not have control over real variables, especially in the medium to long-run. Of course, this is the position taken by orthodox (*i.e.*, pre-Keynesian) macroeconomic theory, and incorporates the idea of the "classical dichotomy" in which nominal variables, like monetary variables, have no long-run impact on real variables, such as employment. There is very little support for this point view, and it may be after all a wrong view.

As Keynes argued, the long run is made up of a series of short runs in which, monetary variables, including those controlled by monetary policy, can have very significant impacts on important real variables, such as employment and economic growth. Frenkel and Ros (2005) and Frenkel and Rapetti (2008) argue that central bank should maintain a moderate inflation rate and should maintain a competitive and stable real exchange rate. They note that the real exchange rate can affect employment, and the economy more generally, through a number of channels: (1) by affecting the level of aggregate demand (*the macroeconomic channel*) (2) by affecting the cost of labor relative to other goods and thereby affecting the amount of labor hired per unit of output (*the labor intensity channel*) and by affecting employment through its impact on investment and economic growth (*the development channel*).

(*e.g*. Frenkel and Ros, pp. 634-637) While the size and even direction of these effects might differ from country to country, in many countries, maintaining a competitive and stable real exchange rate is likely to have a positive employment impact overall. (Frenkel and Rapetti, 2008).

From the point of view of monetary policy, the challenge is how to design a policy to maintain a stable and competitive real exchange rate (SCRER) so that it does not get undermined by massive speculative capital flows.

Financial Policy and Structure

Policies like those described above historically have been quite common in both the currently developed and the developing countries (Epstein, 2007). Over the years, central banks have been seen as agents of economic development, not just agents of economic stabilization. And while sometimes central banks have failed quite spectacularly in this mission, there have been other important success stories, including important periods in the U.S., U.K., France, Germany, Japan, South Korea and India, to name just a few examples. In continental Europe, the banking system, often directed by the central bank in conjunction with the ministry of finance, helped to mobilize and direct credit for industrial development. Even in the U.S. and U.K. these policies were used to direct policy to promote social sectors such as housing; in the U.S. and the U.K., central bank policy and regulations were used to promote the financial sector as well.

As for developing countries, Alice Amsden describes the key role that investment banks played in the industrialization success stories such as South Korea, Taiwan, Malaysia, Brazil, Argentina and others, in mobilizing and directing savings to key industrial sectors, and in particular, those specializing in exports. (Amsden, 2001). In many of these cases, central banks were a key part of the government apparatus that played a supporting role by maintaining low interest rates, maintaining capital controls to help stabilize exchange rates at competitive levels, and sometimes engaging in direct lending for preferred purposes.

In some countries, engaging in these developmental roles, using a wide variety of instruments, was seen as a key part of the central bank's mission. After the Second World War, there was a major transformation of central banking in the developing world. In many respects, these changes paralleled those in the developed world just described. But in developing countries, central banks were much more emphatically *agents of economic development* then in many richer countries (Epstein, 2007a). Hence, we need to re-engage central banks, finance ministries, and the financial system generally in playing a much more active role in generating employment, promoting development, and maintaining financial stability.

This is especially true now that we have seen again the severe dangers of lightly regulated private finance. But even before the most recent disastrous crisis, the promise of financial liberalization for development in poor countries was lacking in supportive evidence.

Still, we must be aware of the failures and excesses of some of these approaches in some countries. As described in Amsden's book, there must be transparent, accountable, and control mechanisms to ensure that these developmentally oriented financial institutions do not contribute to macroeconomic instability, inefficiency or financial or political corruption.

Policy Recommendation on the Role of the Financial Sector

In order to promote more and better employment, poverty reduction and reduce financial instability, other financial policies are required. To achieve these outcomes, financial policies have to serve both general purposes, as well as achieve outcomes quite specific to the needs of particular countries and regions. More specifically, the financial sector can play an important and productive role in promoting employment, growth and poverty reduction through the following channels. It can:

- ✰ Mobilize savings that can be used for productive investment and employment creation;
- ✰ Create credit for employment generation and poverty reduction at modest and stable real interest rates;
- ✰ Allocate credit for employment generation and help the poor to build assets, including in agriculture and in small- and medium-sized enterprises and in housing;
- ✰ Provide patient (long-term) credit for productivity-enhancing innovation and investment;
- ✰ Provide financing for public investment to provide for employment generation and productivity enhancement;
- ✰ Help to allocate risks to those who can most easily and efficiently bear those risks;
- ✰ Help to stabilize the economy by reducing vulnerability to financial crises, pro-cyclical movements in finance, and by helping to maintain moderate rates of inflation;
- ✰ Help the poor by providing basic financial and banking services.
- ✰ Increase domestic investments, and thus potential self-reliance

In order, to promote more and better employment, poverty reduction and reduce financial instability, other financial policies are required. To achieve these outcomes, financial policies have to serve both general purposes, as well as achieve outcomes quite specific to the needs of particular countries and regions. More specifically, the financial sector can play an important and productive role in promoting employment growth and poverty reduction through the following channels. It can:

- ✰ Mobilize savings that can be used for productive investment and employment creation;
- ✰ Create credit for employment generation and poverty reduction at modest and stable real interest rates;
- ✰ Allocate credit for employment generation and help the poor to build assets, including in agriculture and in small- and medium-sized enterprises and in housing;

- Provide patient (long-term) credit for productivity-enhancing innovation and investment;
- Provide financing for public investment to provide for employment generation and productivity enhancement;
- Help to allocate risks to those who can most easily and efficiently bear those risks;
- Help to stabilize the economy by reducing vulnerability to financial crises, pro-cyclical movements in finance, and by helping to maintain moderate rates of inflation;
- Help the poor by providing basic financial and banking services.
- Increase domestic investments, and thus potential self-reliance

Central Bank Policy for Employment Creation

One reason that "inflation-focused monetary policy" has gained so many adherents is the common perception that there is no viable alternative monetary policy that can improve growth and employment prospects. There are three main factors accounting for this perception. First, many economists believe the Pre-Keynesian natural rate (or, alternatively, "Non-Accelerating Inflation Rate of Unemployment (NAIRU)) view of the labor market that claims that, left to their own devices, market forces will automatically bring the economy to full employment and, furthermore, any attempt to reduce unemployment beyond that will only result in ever worsening inflation. However, there is substantial evidence that the NAIRU theory is not empirically well based. The natural rate or NAIRU, if it exists, does not seem to be constant; importantly, it seems to be affected by macroeconomic policy itself. In some countries its effects are asymmetric, with increases in unemployment reducing inflation, but reductions in employment not increasing inflation; and it no longer even seems central to the work of mainstream economics (see Eisner, 1997; Baker, 2000; Ball and Sheridan, 2001; Ball and Mankiw, 2002; Pollin, 2005; Hall, 2005).

Second, in an internationally financially integrated economy with high levels of international capital flows, monetary policy can be extremely challenging. In particular it might be very difficult to gear monetary policy by targeting monetary aggregates, or by pegging an exchange rate along with trying to promote employment growth. This is often seen as the so called "trilemma" which says that central banks can only have two out of three of the following: open capital markets, a fixed exchange rate, and an autonomous monetary policy geared toward domestic goals. While this so-called "trilemma" is not strictly true as a theoretical matter, in practice it does raise serious issues of monetary management (Frenkel and Taylor, 2006). From my perspective, the real crux of the problem turns out to be one leg of this 'tri-lemma", namely the fact that orthodox economists, by and large, have taken for granted that eliminating capital controls is the best policy, and that virtually complete financial liberalization with respect to the foreign sector is the optimal policy. Yet recent evidence amply shows that open capital markets can create very costly problems for developing countries and that many successful developing countries have used a variety of capital management techniques to manage these flows in order, among other things, to help them escape this so-called "trilemma". (See below for further discussion of capital management techniques.) Third, few economists have developed and proposed concrete alternatives to inflation targeting monetary policy in the current context so those searching for alternatives have trouble finding models. However, economists are now in the process of rectifying this problem and I report on these efforts in this section.

The main point of this section is this: there *are* viable alternatives to inflation-focused monetary policy (including inflation targeting), alternatives that can promote more and better employment and poverty reduction. Moreover, these alternatives are also responsive to the needs to keep inflation at a moderate level and to maintain an exchange rate that is not excessively volatile: in short, these alternatives are responsive to stabilization needs as well as developmental needs. The Real Targeting Approach to Central Banking *Advantages of the Real Targeting Framework* This *real targeting* framework has a number of important advantages:

1. First and foremost, it places front and centre the economic variables that have the most immediate and clearest association with social welfare. In the case of implementing a strict "targeting" framework, the central bank will be forced to identify this target and then reach it, and if it doesn't do so, both explain why it failed and how it will improve in the next period.
2. Given the public pressure to reach this target, the central bank will have significant incentives to invest in research and other activities, to improve its understanding and tools to reach this real target.
3. Given that it will need to reach this target amid other constraints, it will need to develop new tools of monetary policy. For example, if a central bank must hit an employment target subject to an inflation and balance of payments constraint, then – in addition to interest rate policy - it might explore asset allocation strategies to encourage banks to increase their lending to high employment generating activities.
4. A real targeting approach lends itself naturally to a more democratic, transparent and accountable central bank policy that serves the genuine needs of the majority of countries' citizens, rather than the minority that typically benefits from the combination of slower growth, low inflation, and high real interest rates.
5. The framework is much more conducive to tailoring monetary policy to the specific needs of different countries. For example, if a country has a particular problem with generating good jobs for women, or more jobs in a particular region of the country, then the real targeting approach can target women's employment or more employment in a specific region (along with more employment generally) and devise instruments to achieve those objectives.

In short, the *real targeting approach* to monetary policy is likely to be more developmentally relevant, flexible, effective and democratically accountable than inflation targeting. An important example of real targeting, developed in more detail below, is *employment targeting*.

How would An Employment Targeting (ET) Framework Operate?

The central bank, in conjunction with the government, would estimate a feasible target range for employment growth, taking into account the rates that are consistent with moderate inflation. Based on the estimate of the relationship between the central banks' policy instrument and employment growth, the central bank will try to achieve its target. Note that many of the problems that arise will be similar to those that arise with inflation targeting. For example, what are the best instruments to use? What is the best way to measure employment growth? What should be done about uncertainty and imperfect information? These are difficult issues. But there is no reason to believe that these issues will be any harder – or easier to deal with–than in the inflation targeting case.

Objections to Employment Targeting

The major objection to employment targeting from mainstream economists is also the least valid: it is the claim that only nominal variables can be affected by monetary policy, at least in the long run. But, as discussed earlier, there is plenty of evidence that central bank policy can have significant impacts on employment and other real variables, such as investment (*e.g.* Ball, 1999). There will be other objections to employment targeting. Many economists will argue that the association between central bank policy and employment growth is simply too loose to base policy on. But Bernanke *et al.*, and other advocates of targeting have admitted that the connection between central bank policy and inflation is also lose and variable. (Bernanke, *et al.*, 1999) There is simply no macroeconomic variable worthy of interest, including inflation that is perfectly controlled by the central bank.

The Indirect Advantages of Employment Targeting

There are several extremely important indirect advantages of the employment targeting framework which themselves will contribute in a crucial way to the framework's success. These indirect advantages involve the accumulation of new knowledge about the connections between monetary policy and employment, and the generation and implementation of new ideas about how to generate more employment and economic growth. Both of these positive outcomes will stem from focusing the attention of the central bank, with its enormous human and financial resources, on the key issue of employment growth.

Making employment or employment growth a target of monetary policy will induce a profound shift in the attitude and the activities of the central bank. It will begin to assign its economists to study the relationships between monetary policy and employment growth; it will study what monetary policy instruments are best used to achieve employment growth; it will organize conferences on employment growth and monetary policy. It will give promotions and more resources to members of its staff that make breakthroughs in the understanding of these connections. It will lead the central bank to link up with others outside the Bank who have knowledge and experience with respect to employment generation and its relationship to financial variables.

The creation of an employment targeting framework will not only lead the central bank to generate more knowledge about the relationship between monetary policy and employment growth. It will lead the bank to design new programs to help it reach it targets. To be sure, lowering interest rates, as I have suggested earlier, could have an important impact on investment and expansion behavior by many firms, large and small. But the central bank will also find that, in order to contribute to more employment generation subject to an inflation constraint, it will need additional instruments of monetary policy.

Targets and Instruments

As Tinbergen argued in his classic piece on policy making, one needs at least as many independent instruments as targets to reach one's objectives. If the central bank were to adopt an employment targeting approach, in which it also had to meet an inflation constraint, then in effect there are up to two targets and only one instrument, the interest rate. In this situation, the Central Bank will undoubtedly search for more instruments with which to reach its targets.

There are a number of useful possibilities. Among the most promising are various mechanisms of credit allocation and capital management techniques.

Credit Allocation Mechanisms

Post-War US, European and Japanese economic policies, central banks have used a variety of tools to help direct credit to preferred sectors to support social and economic goals. These tools include asset based reserve requirements, loan guarantees, support for pooling and underwriting small loans, utilizing the discount window in support of employment generating investments (see Pollin, *et al.*, 2006 and 1994, Epstein and Heintz, 2006, Palley, 2003, and my treatment below for more discussion of these approaches). Asset based reserve requirements are a system in which lending institutions are required to hold low-interest or zero interest yielding reserves as a proportion of their earning assets. Hence, banks loose money in proportion to the size of the reserves they need to hold. To make this part of a directed lending system, the central bank or government would assign lower requirements to preferred assets, such as employment generating activities, thereby creating an incentive for banks to hold these types of assets relative to less desirable assets. Another possible instrument would be to create a loan guarantee program for banks that lend to cooperatives and small businesses that generate good employment opportunities. Another example is to provide liquidity and risk sharing institutions for loans to businesses which have promise to generate employment but which now have restricted access to financial markets. An example of this type of policy would be to provide financial and administrative support for asset backed securities which would bundle loans for employment generation, and sell them as securities on the open market, much as Ginnie Mae and Fannie Mae do in the United States housing market.

As a final example, the Reserve Bank could open a special discount window facility, in which it would offer credit, guarantee, or discount facilities to institutions which are on lending to firms and cooperatives which have promising employment generating opportunities. Of course, these types of policies can lead to abuse and corruption. It is crucial that monitoring and control systems be established to minimize these costs. As Amsden argues (2001), these control and monitoring systems were crucial to the success of newly industrializing countries such as South Korea and Taiwan that used credit allocation techniques to support their industrial policies. In the cases that Amsden describes, firm's success in generating exports was the key monitoring variable. Firms that did not succeed would be penalized.

Employment Targeting Through A Gender Lens

Among the great advantages of combining credit allocation mechanisms with the employment targeting approach is that it provides the policy flexibility necessary to succeed in varying institutional contexts. For example, where female waged employment is important for social welfare, using a gender lens will be crucial for designing effective macroeconomic policy in general and employment targeting monetary policy in particular. In these contexts, the central bank could design asset based reserve requirements to generate more female employment; alternatively, the central bank could give preferential access to the discount window for financial institutions investing in or lending to organizations which will generate more, better, female employment. Indeed, even in the absence of a prior understanding of the importance of women's wage employment, in these situations, an employment targeting central bank will likely discover on its own the importance of using a gender lens in macroeconomic policy if it is going to achieve its employment generating targets. This is an excellent example of the beauty of implementing an employment targeting approach: it will force the central bank to learn much more about the nature of the labour force and how to increase gainful employment which, in many societies, will mean that it will learn much more about the female labor force.

Capital Management Techniques

As I discussed earlier with respect to the so-called *tri-lemma*, governments may need policy space from the vagaries of international capital markets to preserve autonomous monetary and other policies along the lines discussed here. In particular, to achieve this policy space they may need capital controls, exchange controls or, more generally, capital management techniques.

Capital and Exchange Controls place taxes or regulations over the buying and/ or selling of international assets or debt (capital controls) or foreign exchange (exchange controls). These have been used widely to manage inflows and outflows of international capital and/or foreign exchange. *Capital management techniques* (CMT's) refer to two complementary (and often overlapping) types of financial policies: policies that govern international private capital flows (capital and exchange controls, as just defined) and those that enforce prudential management of domestic financial institutions. Certain types of prudential financial regulations actually function as a type of capital control; moreover, capital controls themselves can function as or *complement* prudential financial regulations. There is often a great deal of synergy between prudential financial regulations and traditional capital controls.

It can be difficult (and sometimes impossible) to draw a firm line between prudential domestic financial regulation and capital controls. For instance, domestic financial regulations that curtail the extent of maturity or locational mismatches may have the effect of influencing the composition of international capital flows to a country, even if those types of regulations are commonly classified as prudential domestic financial regulations and not as capital controls. Hence, for the IMF and others to argue that prudential management is "good" and capital controls are "bad" is to draw a distinction for ideological, rather than sound policy purposes.

Policymakers use Capital Management Techniques (CMTs) to achieve some or all of the following four objectives–to promote financial stability; to encourage desirable investment and financing arrangements; to enhance policy autonomy; and to enhance democracy. CMTs can promote financial stability through their ability to reduce currency, flight, fragility and/or contagion risks, thereby reducing the potential for financial crisis and attendant economic and social devastation. They can also influence the composition of the economy's aggregate investment portfolio, and can influence the financing arrangements that underpin these investments. Moreover, CMTs can promote desirable types of investment and financing strategies by rewarding investors and borrowers for engaging in them. Desirable types of investment are those that create employment, improve living standards, promote greater income equality, technology transfer, learning by doing and/or long-term growth; desirable types of financing are those that are long-term, stable and sustainable. Capital management can discourage less socially useful types of investment and financing strategies by increasing their cost or precluding them altogether.

CMTs can enhance policy autonomy in a number of ways: they can reduce the severity of currency risk, and can thereby allow authorities to protect a currency peg; they can create space for the government and/or the central bank to pursue growth-promoting and/or reflationary macroeconomic policies by neutralizing the threat of capital flight; by reducing the risk of financial crisis in the first place, CMTs can reduce the likelihood that governments may be compelled to use contractionary macro- and micro-economic and social policy to attract foreign investment back to the country or as a precondition for IMF assistance; finally, CMTs can reduce the specter of foreign control or ownership of domestic resources. It follows from the above that capital management can enhance democracy by reducing the potential for speculators and external actors to exercise undue influence over domestic decision making directly or indirectly (via the threat of capital flight). CMTs can reduce the veto power of the financial community and the IMF, and create space for the interests of other groups (such as advocates for the poor) to play a role in the design of policy. They can thus be said to enhance democracy because they create the opportunity for pluralism in policy design.

Regimes of capital management take diverse forms and are multi-faceted. Importantly, capital management techniques can be static or dynamic. Static management techniques are those that authorities do not modify in response to changes in circumstances. Capital management techniques can also be dynamic, meaning that they can be activated or adjusted *Other Possible Real Targets for Monetary Policy.*

In addition to a direct employment target for monetary policy, there are other real or intermediate targets that could be desirable candidates for central banks wanting to better balance developmental goals with inflation stabilization. Here I will briefly mention three others: first is real GDP growth, second is investment targeting, and third, is a *stable and competitive real exchange rates* (*SCRER*). The latter is widely proposed in the case studies presented below.

Real GDP Growth (Economic Growth)

Economic growth is a natural alternative target for central banks, since it is generally considered a broad measure of economic performance and is widely thought to be influenced, at least in the short to medium term, by macroeconomic policy, including monetary policy. Moreover, many other important variables, such as employment and investment are broadly affected by the rate of economic growth. Its breadth and generality might also make it less amenable to monetary and credit policy than more narrow measures such as employment and investment. As a broad measure, it might also be less directly linked to specific outcomes with high social priority, such as poverty reduction than employment.

Investment

For some countries, employment generation is not the most immediate macroeconomic problem. Rather, productivity growth is a more important concern. In this case, investment, rather than employment, might be a more appropriate target. If the central bank were to adopt an investment target subject to an inflation constraint, it could do so in a targeting framework. In this case, the central bank would adopt target ranges for investment and would estimate the relationship between central bank policy and investment. It would also estimate the relationship between policy, investment and inflation and other important variables, such as exchange rate instability. It would choose a target range for inflation and assess whether the investment goal and the inflation goal are consistent. Just as in the inflation targeting framework, decisions will have to be made as to the proper index of investment to use. Based on the estimate of the relationship between the central banks' policy instrument, the central bank will try to achieve its target. Here too the problems that arise will be similar to those that arise in the inflation targeting approach. What is the best instrument to use? Most central banks use short term interest rates as their policy tool. But other variables are likely to be more directly related to investment, including long term interest rates and credit availability.

Therefore, Central Banks may find that other instruments, such as open market operations directed at long-term debt securities, directed lending, or incentives, such as asset based reserve requirements to induce banks to lend longer term, are more effective. (See the discussion on South Africa below).

A Stable and Competitive Real Exchange Rate (SCRER)

As will be clear from our discussion below, a number of the authors in the PERI/Bilkent research group have chosen a *stable and competitive real exchange rate* as an important real target for monetary policy. This is because in the countries they are studying, international competitiveness has a large impact on employment, productivity growth and economic growth. Moreover, they argue that targeting the real exchange rate is feasible for central banks, and can relatively easily accommodate a focus on stabilizing inflation, which for many Latin American countries, is a very important objective. More details on the SCRER and how it might operate are provided below.

Alternatives to Inflation Targeting: Variations on a Theme

Modest but Useful Adjustments to the Inflation Targeting Regime Some of the country studies in the PERI/Bilkent project proposed only modest changes to the inflation targeting regime. In the case of Mexico, for example, the authors argue that the existing inflation targeting regime has allowed for more flexible monetary policy than had occurred under regimes with strict monetary targets or strict exchange rate targets (Galindo and Ros, 2006). As a first step, they suggest modifying the IT framework to make it somewhat more employment friendly. More concretely, Galindo and Ros find that monetary policy in Mexico has been asymmetric with respect to exchange rate movements –tightening when exchange rates depreciated, but NOT loosening when exchange rates appreciated. This has lent a bias in favor of an over-valued exchange rate in Mexico. In its place, Galindo and Ros propose a "neutral" monetary policy so that the central bank of Mexico responds symmetrically to exchange rate movements and thereby avoid the bias toward over-valuation without fundamentally changing the inflation targeting framework. In his study of Brazil, Nelson Barbosa-Filho also proposed extending the inflation targeting framework, but, as discussed shortly, in a more dramatic way. Writes Barbosa-Filho:

"...because of Brazil's past experience with high inflation, the best policy is to continue to target inflation while the economy moves to a more stable macroeconomic situation. So far, the great gain from inflation targeting has been the increase in the transparency and accountability of monetary policy in Brazil." (Barbosa-Filho, 2005). But he goes on to say,

"The crucial question is not to eliminate inflation targeting, but actually make it compatible with fast income growth and a stable public and foreign finance". (ibid.) As discussed in the next section, in order to do that, Barbosa-Filho joins a number of the country case study authors in proposing a monetary policy to maintain a stable and competitive real exchange rate (SCRER) which, they argue, will have a number of significant benefits for many of these economies and their peoples.

A Competitive and Stable Real Exchange rate

As just indicated, a number of authors, following the lead of Frenkel and Taylor (2005), Frenkel and Ros (2006) and Frenkel and Rapetti (2006) argue that central bank should maintain a moderate inflation rate AND should maintain a competitive and stable real exchange rate. They note that the real exchange rate can affect employment, and the economy more generally, through a number of channels: (1) By affecting the level of aggregate demand (*the macroeconomic channel*)

(2) By affecting the cost of labor relative to other goods and thereby affecting the amount of labor hired per unit of output (*the labor intensity channel*)and by affecting employment through its impact on investment and economic growth (*the development channel*). (*e.g.* Frenkel and Ros, pp. 634-637) While the size and even direction of these channel effects might differ from country to country, in many places, including Latin America, maintaining a competitive and stable real exchange rate is likely to have a positive employment impact though some combination of these effects. For example, Frenkel and Ros find that in Latin America, in most of the Latin American countries experiencing increases in unemployment in the 1990's, these economies were characterized by significant appreciations of their real exchange rate.

A Three Step Framework

Objectives of the Framework

1. The objectives of the framework provided below are threefold. It aims to help analysts understand first, how the sectoral pattern of growth affects its poverty impact and second, whether employment-intensive or productivity-intensive growth is more important. Finally, the framework aims to shed light on the broader policy and institutional environment which might facilitate the poor's participation in growth.
2. Alongside gaps in relevant data in many developing countries, the difficulty of assigning causality in the relationship between growth, employment and poverty reduction is probably a major reason for deficiencies in the current research agenda. A thorough analysis of the mechanisms by which the sectoral patterns of growth impact poverty would require a complex general equilibrium model. Such models risk stretching data beyond their limits and also pose immense challenges to analytical capacity.
3. Recognizing that the demands of general equilibrium models are often prohibitive in the context of limited data and capacity, the framework offered here is more modest. It provides analysts and policymakers with tools to observe correlations between employment generation and productivity growth, poverty reduction, and the broader institutional and policy environment. While correlations are of course not an indication of causality, this framework offers a pragmatic and feasible response to the absence of a thorough analysis of even correlative relationships in many countries. It can provide empirical input into debates on how the quantity and quality of jobs are linked to poverty. And it may help to stimulate discussion on the appropriate policies for enabling the poor's full participation in economic growth.

The Three Step Framework

Step 1: Profiling Growth To what extent is growth associated with changes in employment, output, or population? Step 2: Linking the Growth Profile to Poverty Reduction To what extent are changes in the employment, output and population rates linked to poverty reduction? Step 3: Identifying Policies and Institutions that

Matter To what extent is a poverty reducing growth profile linked to the broader policy and institutional environment?

The framework provides three stages of analysis. First, it decomposes economic growth into changes in both the quantity and quality of jobs. Second, it ascertains whether improvements in employment or productivity are most readily associated with poverty reduction, and whether this pattern changes by sector. Third, it highlights the broader policy and institutional environment which is associated with a poverty reducing pattern of growth.

Step 1: Profiling Growth

The first step in understanding how sectoral growth affects poverty via its impact on the quantity and quality of jobs is to find a suitable methodology for profiling growth.

Analysis of the relationship between growth and employment has traditionally focused on the employment elasticity of growth – that is, the percentage change in employment in response to 1 percentage change in output. This measure has three notable shortcomings. First, it says nothing about the actual extent of job creation since a country that grew by 1% and enjoyed a 1% increase in employment would have the same employment elasticity rate as a country which had a 10% growth rate accompanied by a 10% increase in employment. Second, the measure does not take demographic changes into account, although growth in population often swamps increases in employment. Third, and most obviously, the employment elasticity of growth says nothing about the quality of new jobs created.

A simple way of understanding how growth is associated with increases in the quantity and quality of jobs is to perform a decomposition of *per capita* GDP growth into three components: productivity changes, employment rate changes and demographic changes. The amount of growth that can be attributed to changes in output per worker can be obtained by calculating the resulting growth in *per capita* value added between two points in time under the hypothetical scenario in which employment rates and the share of the working age population remain constant, but output per worker has changed as observed. Similarly, the amount of growth attributed to changes in employment or the share of the working age population can be obtained by calculating *per capita* value added when the other two components are held constant.

The results of this growth decomposition, using the example of Nicaragua over a four year period (2001-2005), during which value added growth totalled 12%. It shows that almost three-quarters (74%) of the increase in growth can be linked to changes in the structure of the population. In other words, had everything else stayed the same, the reduction in the number of dependents per working age person would have generated a growth rate of 5.3% (or 74% of 7.12). Changes in employment were also important: accounting for over half (51%), of the observed growth. This means that if productivity and the number of dependents per working age member had both remained constant, the higher rate of employment would have generated a GDP growth rate of 3.6% (or 51% of 7.12).

Unfortunately, changes in productivity acted in the opposite direction. Indeed, had population and employment levels remained constant, shrinking productivity would have had a negative impact equivalent to more than a quarter (26%) of the actual observed growth rate. Put differently, had productivity not changed, observed growth would have been close to 9%, but decreases in productivity meant that growth was 7.12%.

12

Globalization and Its Conceptual Usages

12.1. Globalization and Its Conceptual Usages

The phenomenon of globalization is a multi-dimensional and multifaceted process that encompasses political, economic, social and cultural dimensions that have been variously explained in different terms and contexts. Viewed from a general perspective, the notion of globalization is broad and diverse. Our analysis in this paper is focused on its challenge for Nigeria's development in the 21st century. It is considered germane however to consider first the multi-dimensionality of its conceptual usages. Due to its multivariate nature, globalization does not lend itself to easy conceptualization; and like other concepts in social sciences, it is not amenable to a single, simple and straight jacket definition, which perhaps explains its various connotations by scholars of different persuasions as internalization, universalization, liberalization, westernization etc. As Jibrin Ibrahim, drawing from James Mittleman, insists, globalization is not a single unified phenomenon but rather a syndrome of processes and activities, which embody a set of ideas and a policy framework organized around the global division of labour and power (Ibrahim, 2002:3). In a similar vein Tade Aina forcefully submits that:

Any meaningful and relevant understanding of globalization in Africa must go beyond the myths and ideologies of globalization to the confrontation with the diverse but actual processes, how they unfold, their relationships with themselves and other social and economic relations and dynamics. Such an understanding must also recognize not only the complex but varied history of the processes being studied but it must reject a monolithic or homogenized under- standing such as that found in the currently neo-liberal confrontation of the subject. (Quoted in Dauda Abubakar, 2001:16). There is therefore no unanimity of opinion on what constitutes globalization; neither is there a consensus concerning the extent of its advancement as a concept. For a thorough understanding of the concept however, there is the need to reflect briefly on the thrust of the discourse which in literature has been classified into three main clusters globalization as implying global culture; globalization as the expression of the global political order; and globalization as depicting global economy. Hence, as Symonides perceptively submits, globalization is generally the process of growing interconnection and interdependence in the modern world. It is generated by growing economic, cultural and political cooperation and links, as well as by the need to respond together to global problems which can be solved only on a planetary scale (Symonides, 1998:28).

Dauda Abubakar pungently argues that globalization entails universalization whereby the object, practices or even values transcends geo-political boundaries, penetrating the hitherto sovereign nation state and impacting the orientation and value system of the people (Abubakar, Op.cit). In a broader perspective, Tade Aina posits that globalization depicts the transformation of the relations between states, institutions, groups and individuals, the universalization of certain practices, identities and structures, and perhaps more significantly, the expression of the global restructuring that has occurred in recent decades in the structure of modern capitalist relations (Aina, 1996). In its most generic and broadest sense, Mike Kwanashie sees globalization as a part of the movement of history as evident in certain forces that appear to push for increasing integration of human activities with emphasis in contemporary world focused more on the economic aspect of the process. It is a process of increased integration of national economies of states with the rest of the international system in order to create a more coherent global economy (Kwanashie, 1999:17). The process has been increasingly propelled by the revolutionary trend in information technology that combines advancement in computing, electronics, and telecommunications which has brought up a highly dynamic process of storing, processing, transmitting and presenting information (Ibid:20). As Nuhu Yaqub copiously observes, globalization refers to the tremendous revolutionary changes that have affected our planet as a result of changes that have also taken place in information and communication technologies - processes that have, cumulatively, led to the villagization of the globe (Yaqub, 2003:45).

Broadly speaking and from the perspective of the UNDP National Human Development:

Globalization can be defined as a multidimensional process of unprecedented rapid and revolutionary growth in the extensiveness and intensity of interconnections on a truly global scale (UNDP: National Human Development Report, 2000/2001). From the foregoing, a fundamental feature of globalization is that the economies of modern nation states have been highly integrated to the extent that the internationalization of trade and economic activities has become imperative. The interdependency is profoundly informed, as Mufana Lipalile observes, by enhanced international trade, sustained financial flows, increased inter-country mobility as well as improved communication arising from the innovative existence of cellular telephones, electronic mail, the internet and digital satellite television (Lipalile, 2001:296).

12.2. Strands of Thought on Globalization

There exists a plethora of works on globalization (see for instance Aina, 1996; Ake, 1995; Appadurai, 2001; Clark, 1999; Nabudere, 2000; Ninsin, 2000; Mittleman, 2000; Rugumamu, 1999; Scholte, 2000; Tandon, 2000; etc.), perspectives on the subject however differ fundamentally depending on the ideological conviction of different scholars. In general, the phenomenon has been conceived from two contrasting paradigms namely globalization as interdependence and globalization as imperialism. Scholars of globalization as interdependence are of liberal persuasion. They see the concept as a framework of complex and growing interdependence among nations. The global socio-political and economic integration is viewed in the context of inter dependencies which has restructured the world into a new and all inclusive social pattern. They associate globalization with economic liberalization as a policy option for the development of the south through a process of free trade, investment and capital flows between countries.

Scholars of globalization as interdependence have consistently maintained that globalization is the rational end point of human development, and that it is capable of impacting positively on the life of state actors that integrate their economies. Francis Fukuyama for instance, in his treatise. The end of history and the last man perceives globalization as universalization of western values. He proposed the celebration of a globalized world and the unabashed victory of political and economic liberalism that is evidenced in the triumph of western ideas and values and in the exhaustion of viable systematic alternatives to western liberalism (Fukuyama, 1992). Severine Rugumamu opines that globalization is not merely a buzzword; rather it is a new paradigm in international economic relations which apparently signals the triumph of capitalism on a truly global scale following the end of the cold war, the collapse of the Soviet system and the dissolution of planned economies, particularly in Eastern Europe (Rugumamu, 1999:3).

Jan Scholte identifies globalization as deterritorialization or as spread of supra-territoriality. He opines that the phenomenon constitutes a transformation in the spatial organization of social relations and transactions (Scholte, 2000). Proponents of globalization as inter-dependency therefore see a better world if nation states would realize and utilize to the maximum the opportunities presented by inter-dependency resulting from globalization. Their belief is hinged on the premise that interdependency has opened up the world, reduced the abuse of human rights, and eradicated, to a large extent, social and economic injustices by national governments.

Advocates of globalization as imperialism on the other hand are mainly of the radical persuasion and political economy genre. While the interdependence school of thought on globalization claims that interdependence is the reality of globalization and that it constitutes a positive development in world affairs, scholars who view globalization as imperialism insist that the phenomenon as it is today represents nothing but capitalism and imperialism. Ali Mazrui calls it the new global imperialism. Scholars that allude to the same position have proclaimed that globalization is a transformatory capitalist project, which can only serve to impoverish the underdeveloped nations on the fringe of the world capitalism. Claude Ake, for instance, sees globalization as a capitalist project that is structured to perpetuate the underdevelopment of Africa and other Third World countries. He construed globalization in terms of profit maximization, and perceptively referred to it as the march of capital across the world, in search of profits; a process that is facilitated by the expansion of multinational corporations, and driven by the technical advances in communication. In his words: Globalization is about growing structural differentiation and functional integration in world economy; it is about growing interdependence across the globes, it is about the nation-state coming from under pressure from the surge of transnational phenomenon; about the emergence of a global mass culture driven by mass advertising and technical advances in mass communication (Ake, 1995).

Alluding to Akes contention, Edwin Madunagu sees globalization as the rapid expansion of capitalism. He believes that:

The rapid expansion, through giant multinational companies, of capitalism to several areas of the world, including areas where it had hitherto been resisted or put in check side by side with this expansion, is the phenomenal development of computer technology, telecommunication and transportation. The latter serve as the main vehicle of the former. Globalization is globalization of capitalism, not the globalization of a neutral economic system or globalization of post-capitalism as the imperial intellectuals and their slaves in the underdeveloped countries would have us believe (Madunagu, 1999:53).

Daniel Nabudere similarly submits that globalization is a capitalist economic project that is propelled by two contradictory movements first, the tendency of economic globalization to create uniformities in the entire world; and second, the tendency of marginalization and fragmentation, which the phenomenon connotes. He summed up that these two processes make it possible for globalization to spread out its risk and losses throughout the global arena (Nabudere, 2000). Thus, can it be contended that globalization appropriates development for one part of the world and underdevelopment for another? How has this been actualized? and what should be done in order to confront the challenges posed? There is no doubt that globalization has impacted differently on the two parts of the global system. While it strengthens the already developed advanced western countries, it marginalizes the peripheral economy of the Third World countries. K. A Ninsin elegantly captures the scenario in his description of globalization as the engine by which the economies of the worlds weaker nations are being opened up and subjected to the hegemony of the developed capitalist economies (Ninsin, 2000:5). Felix Oriakhi alludes to the same position when he states globalization super-imposes the values of the North on the South by accelerating the successful penetration of capitalism and other bourgeois ethos into the nooks and corners of Asia, Latin America and Africa and the erstwhile non-capitalist states of Eastern Europe (Oriakhi, 2001:24). In a seminal work, Esko Toyo insists that globalization is an alternative imperialist policy deliberately designed to maintain their structure of impoverishment on their satellite states (Toyo, 2000).

The two paradigms - globalization as interdependence and globalization as imperialism-generally reflect the deep ideological and political predilections of the different scholars. What is of more concern and relevance however is that globalization, from which ever perspective it is construed, has its merits and shortcomings; and as Bill Clinton, former US president, contends, it is a fact and not a policy option (Quoted by Lipalile, 2001: 295). Cuban president, Fidel Castros position is equally germane. Globalization, to him, portends a law of history and the consequence of the development of the productive forces, or in simple terms, the development of science and technology (Ibid). Hence, inspite of its adverse effects, it is in the interest of a nation to harness whatever the benefits of globalization for upward national mobility.

12.3. Forces of Globalization

Three main powerful forces that propel globalization in contemporary global environment can be discerned, namely, technological revolution, economic liberalization and democratic system of governance. (Usman, 1999:45 63; Kwanashie, 1999: 15-27).

I. Technological Revolution

That the world is currently experiencing phenomenal changes in social, political, economic and technological spheres cannot be disputed. One fundamental instrument for this increasing transformation is technology, especially computer technology, and the evolution of low cost, global communications system which constitute major challenges that will dominate and fundamentally shape developments in the 21 st century, particularly in the economic and financial sectors. (Usman, 1999:48). The revolution in computer technology, has led to the advancement of information technology, which has in turn enhanced the level of information transmission and business transactions across the global system. Indeed, the revolutionary changes that have occurred in computer technology since the latter part of the 20 th century have brought about tremendous improvements in all facets of human endeavour. Today, the use of computers has increasingly become quite commonplace in pure scientific research, social science and especially in managerial decisions. The significance of the strong force of technology can best be illustrated by reference to the financial system. Globalization and information technology have thrown up formidable challenges for national economies, especially the financial system, by reducing the world further into a global village, and as well by providing enormous information through a wide range of inter-connectivity. The interconnectivity (network) of computers has given rise to the development of Internet, which constitutes the largest network and largest reservoir of all types of information in the global system (Ibid).

Furthermore, information technology in particular has combined progress in electronics, telecommunications and computing to bring about a highly dynamic process of storing, transmitting, processing and presentation of information. This has led to increasing capacity for new and efficient responses to antiquated problems. For instance, more efficient production processes are now possible and countries with the necessary capacity are embracing and adopting them in a bid to maintain a competitive edge within the changing world environment (Kwanashie, 1999:20). Information technology has equally provided tremendous opportunity for the exploitation of economies of scale, making for rapid growth and conferring comparative advantage to those that have access to it. Consequently, it is increasingly promoting the internationalization of productivity and markets, which is crucial to the globalization process (Ibid).

II. Economic Liberalization

One dynamic and fundamental force in contemporary globalization process is economic liberalization, which has been embraced by virtually all countries and major international institutions within the global system. Economic liberalization refers to the process of achieving unobstructed economic activities. It seeks to remove all obstacles to trade, production and investment; emphasizes freedom of economic activities and dominance of private enterprises; and aims ultimately at the divorce of the state from the economy (Onyekpe, 2001:52). At the global level, economic liberalization attempts to make all economies fully open for free inter-penetration and inter-state access. Issues relating to the forces of economic liberalization are generally more complicated and not as obvious as in technology. In pursuit of economic liberalization, the global system is polarized into various groups. In one group are the industrialized (G8) countries that work intimately with the world financial (IMF, World Bank) and Trade (WTO) institutions, and pursue a broad and ambitious agenda that attempt to build international capitalism on the foundations of open world trade and capital flows, privatization, balanced budget, freeing up of exchange controls and similar deregulation and liberalization measures (Usman, Op. cit: 52).

In another group are the Asian- Tigers who have, through economic liberalization process, achieved unprecedented growth in their fragile economies. The countries, in addition to economic liberalization, have used other policy instruments to expand Foreign Direct Investment (FDI) and attain higher economic growth. These include developing a strong production base; opening up of new investment areas; as well as designing and implementing sound macro-economic policies. They have also created a conducive climate and stable political and economic environment to attract foreign investments. These policy measures have enhanced the expansion of FDI in the countries, and led to their rapid growth and development. Other groups of countries that have achieved momentous feat as a result of economic liberalization include the transition economies of the Eastern Europe as well as the Latin American countries. Also included are the African States most of which have suffered tremendously from extreme poverty and lack of policy focus. Thus, it is obvious that economic liberalization is now a common feature among nation states.

It can therefore be rightly concluded that no country in contemporary world can really be an island unto itself, either due to its vibrant economic strength *e.g.* United States or because it decides to close its doors to the outside world like China successfully did in the past. The critical message of globalization in this context is that in the existing moment of integration of global markets, Nigeria and other developing nations have little choice but to try and join the globalization train despite of their disadvantaged position in the process.

III. Democratic System of Governance

The third major force of globalization in contemporary world is the general acceptance of democratic forms of governance. In modern political system, democracy is the preferred system of governance. This is in view of its relationship to good governance with its inherent features including rule of law, democratic participation, probity, integrity, and transparency. It is also preferred because it posits and insists that power springs from, and therefore belongs to the people and that those who exercise power should use it in the interest of the people, or at least, the majority of the people (Yusuf, 1994: 113). In addition, democracy demands unequivocally that while the majority should rule and even in most cases, have its way, the minority must be able to have its say, and its say must be listened to while its rights must be respected and protected. Preference for democracy also springs from the fact that it demands that people should be availed equal opportunities to participate in the administrative and decision making machinery of the state; and that there should exist a free, fair and independent judiciary; equality of all before the law; supremacy of the law; and fundamental human rights for the people (Ibid).

While there has previously been virulent anti-democratic tendencies and suppression of democratic rights of the people, the pendulum is contemporaneously swinging in favour of democracy and democratic governance. Democracy, although varied in form, has in contemporary global system, become widely accepted as the form of governance that advances the interests and aspirations of the majority of the people. This is underscored by the failure of various authoritarian regimes including the erstwhile military and civilian dictatorships in many parts of the world, as well as the collapse of the communist system that provided for usurpation of power by a small clique of bureaucrats over every one else in the society (Usman, Op. cit: 54). The manner of rejecting and isolating non-democratic regimes globally also attests to the changing fortune of democracy as a formidable force and most acceptable system of governance in the global system.

12.3. Challenges for Nigeria

Nigeria, at independence in 1960, was largely a producer and net exporter of primary products. The six major agricultural products then were cocoa, rubber, palm oil, groundnut, cotton and palm kernel. Although there existed mining and quarrying activities, these were of negligible percentage and never counted for the economy as a whole. In other words, agricultural produce and raw materials constituted the sole foreign exchange earner for the country. Specifically, the Nigerian state as an exporter of agricultural goods had 69.4% of its total GDP for the year 1963/64 comprising the six aforementioned agricultural commodities (Olaloku, 1979:8).

The trend of having agriculture as the main foreign exchange earner for Nigeria stopped in the early 1970s when the country was suddenly awash with petrol dollars arising from the quadruple increase in the price of oil in the world market. From 1972 onwards, oil gained ascendancy over all other commodities as the largest contributor to the GDP, and also as a major foreign exchange earner. There was substantial increase in oil production accompanied by a sharp increase in the global market price of high-grade crude oil from low price of $3.8 per barrel in October 1973 to a skyrocketing price of $14.7 per barrel in January 1974. This trend continued till 1981 when the price of crude oil attained a high level of $38.77 per barrel. Within the same period, total revenue from oil rose correspondingly from N1billion to N4billion while external reserves increased from N180 million to N3.7 billion in 1975 (Osaghae, 1998:96).

The increase in national wealth resulted in the government of the day embarking on rapid expansion of the public sector and squandering of the wealth on expanding distributive instead of productive capacity; and on increased dependence on external goods and inputs (Ibid). By 1978, oil (Petroleum and associated components) composed a total of 89.1% of Nigeria's export as against agriculture, which had plummeted in its contribution to export to 6.8% in the same year. The fact of the matter is that Nigeria's commodity pattern has, since the advent of oil, been a mono-cultural one with the product being the only one the country depends upon for its foreign exchange earning - a situation that has constrained the pace of its developmental efforts. Thus, it can be unequivocally asserted that Nigeria's development impotence cannot be solely attributed to colonialism, neo-colonialism and imperialism as often perceptively advanced by scholars of radical persuasion; but also to its own weak domestic economic structure. There is therefore the necessity for the country to diversify its economic base in order to confront the challenges of contemporary globalization process and remain relevant in the scheme of world events. The economies of South East Asian Countries especially the Asian Tigers namely Taiwan, Hong Kong, Singapore and South Korea as well as Thailand, Malaysia and Vietnam are successful stories of how diversification has enhanced their economic developments. The success recorded by this Newly Industrialized Countries (NIC) can be duplicated by Nigeria in its current efforts toward economic transformation.

For instance, as a means of liberalizing its economy, revitalizing its financial sector and generating foreign exchange earnings, Thailand placed top priority on exports. Its economy was liberalized thus encouraging foreign investors. The country's manufacturing sector has emerged as a formidable engine of development in the kingdom with traditional exports in the agricultural sector performing quite successfully. In 1996, Thailand exported about US$9 billion worth of goods to the European Union alone, and by 1997, the country's exports had become highly competitively priced for foreign buyers (Akinbobola, 1999:65). Similarly, in the manufacturing sector, there has been a phenomenal increase in the exports categories. A dramatic development has been the automotive industry in which several big automobile makers have set up production lines in Thailand, thus making the country to become a regional leader in the production and exportation of automotive. In 1996, automotive exports to the European Union alone amounted to US$205million signifying an increase of 38% over the preceding year (The Economist, cited by Akinbobola: 66).

It is interesting to note however that in spite of Thailand's advent of high technology, the country has remained deeply committed to its agricultural base and has indeed emerged as one of the worlds leaders in agro business. In general, Thailand has confronted the challenge of globalization by liberalizing its economy, embracing internationalization of capital and opening new markets across the length and breadth of the world. Hence, Thais products (manufacturing and agricultural) are found all over American, European, Asian and African countries (Ibid). Nigeria can tremendously benefit from Thailand's experience. The Nigerian economy must be completely diversified rather than remaining mono-cultured.

Furthermore, there is the need to ensure a free competitive economy. Taiwan represents one of those Asian nations that have transited from government guidance to a free competitive economy, and thus liberalizing its economy for foreign investors. Today, there exists in Taiwan, foreign investments in construction, power generation, oil refining, real estate, telecommunications and gas stations. In concrete terms, foreign investment has been quite staggering. For instance, United States accumulated investment in Taiwan amount to US$7 billion representing about 30% of total approved foreign investment in the country. Taiwan has huge foreign investments in South East Asian countries and all over the globe. The country also has a large chunk of human resources ready to be utilized by foreign firms. All these have been made possible because of the country's dynamic efforts towards a free competitive economic structure.

The challenges posed by globalization for Nigeria are multifarious. These are considered in the same manner that we articulated the forces that propel it. First, for the country to be fully integrated into the world economy and in order to harness the benefits of such integration, it must embark on serious technological revolution. As Shamsudden Usman pungently remarks, the pervasiveness of technology is such that a country ignores it at its own peril (Usman, Op cit: 56). In order to ensure technological revolution, the country's technological base must be developed. Science and technology should be made a formidable part of the key strategic area of its developmental effort. The need to fundamentally transform the nation's educational, health, agriculture and industrial development has therefore become an imperative demand. Due attention must equally be focused on information technology and telecommunication. In contemporary global economy, information technology and the ability to use it are very critical. While computers and other electronics gadgets are already in place and do not need to be re-invented, there is the need to ensure that the enabling environment for their uses is provided. Hence, adequate basic infrastructures such as power supply and telecommunications must exist regularly and uninterruptedly. Similarly, industrialization, including manufacturing and fabrication should be re-invigorated, while agriculture should be re-positioned with mechanized farming fully embraced to enhance agricultural productivity.

The second challenge relates to economic liberalization. The liberalization policy of the Asian Tigers tremendously enhanced their development. This entails liberalizing the economy, internationalizing of capital, opening new markets and attracting new investments. This poses a great challenge to Nigeria. We agree with

Usman that the Nigerian economy must not only be diversified, but also built on sound economic policies including those that will necessarily ensure increased domestic savings; continued reform of the domestic financial sector; opening up to foreign capital inflows, while simultaneously protecting the country from the huge destabilizing effects of short-term, speculative capital inflows; and together with other developing countries, continue to champion the necessity for the reform of the global financial system that ensures shared prosperity and a greater inducement to the development of the weaker countries (Ibid:58). The heavy dependence of the country on crude oil exports has unprecedentedly exposed the economy to the boom-and-bust cycles and the concomitant unstable and unpredictable volume of revenue receivable by the government (Yaqub, 2003:41). For instance, Nigeria's total export receipts from goods, services and transfers dropped from N10,899.6 million in 1979 to N7,884.2 million in 1983. The countrys import, at the same time increased from N9,890.1 million to N11.022.1million during the same period (CBN Economic and Financial Review 1981-1986). This also affects contributions to the country's Gross Domestic Product (GDP). For instance as at 1999, oil and gas contributed 36.5% of the GDP, while agriculture (including livestock) accounted for 32.8%. In the same period, wholesale and retail trade contributed 16.6%, while manufacturing accounted for only 5.5% of the GDP (Anya, 2001:15). There is therefore the need for the country to diversify the economy and focus on non-oil sector, particularly agricultural and mineral resources.

In addition, the country should maintain a healthy investment climate that can be cashed on by foreign investors. It should also pursue efficient and effective economic management of the country's resources so as to raise the people's standard of living and overall economic development of the nation. It must be stressed however that in its bid to liberalize the economy, the unbridled activities of the Multinational Corporations in the country should be closely monitored and controlled so that the country will not be reaped off by foreign capital. The fact that globalization entails opening up does not signify that the economy should be completely left uncontrolled. Democracy constitutes the third challenge that globalization poses for Nigeria. Democracy has become an acceptable form of governance in the world system. Indeed, no authoritarian or dictatorial regime is fashionable any longer in the global environment. As observed by the UNDP National Human Development Report: The end of the cold war provided a historic opportunity for a world-wide liberal democratic revolution. In view of the western industrialized market economies, the collapse of authoritarianism and socialist central planning has revalidated the claim of liberal democracy, the doctrine of individual freedom and popular democracy, as an ideology of potentially universal validity (UNDP, National Human Development Report for Nigeria, 2000/2001, p.2). Nigeria regained its civil rule in 1999 after about fifteen years of uninterrupted military dictatorship. The persuasive lack of democratic system of governance for several years had significantly deteriorated the developmental pace of the country. Following the restoration of civilian government therefore, there is the need for democracy to be fully entrenched and sustained. As stressed in the National Human Development Report, there must be a strong and vibrant civil society, good governance, effective dynamic relationship between people and the government and entrusting the management of the state in the hands of men and women of high technical competence and integrity (Ibid, p. 42). The country must be truly committed to, and demonstrate the capacity for good governance as a bedrock for a durable democracy.

In this book, it has been demonstrated that globalization as a phenomenon has assumed a new phase in contemporary global political economy, and that Nigeria must equip and package itself effectively to confront its challenges in the 21st century. The paper also articulates the view that globalization in its current dimension is characterized by three distinct but interrelated features: the integration of the economic systems of nation states into global economy; the primacy and supremacy of international markets and international competitiveness; and the phenomenal rise in the internationalization of labour, capital and portfolio investments. The book critically examined the three forces that propelled globalization namely technological revolution, economic liberalization and democratic governmental system. It submits that the new thrust of globalization makes it imperative for nation states to understand the intricacies of the phenomenon so as to enable them devise strategic ways to harness its advantages. The challenges of globalization must be faced by developing nations for them to be relevant in the global scheme of things. For Nigeria, in particular, the challenges are enormous but not insurmountable.

Nigeria cannot afford to remain aloof and be completely isolated from the global economic interplay. The country should make conscious effort to transform the economy so as to fully harness the benefits of globalization. The various endogenous and exogenous constraints that have continued to undermine macro economic stability in the country should be seriously tackled. The critical sectors of the economy that constitute the basis for sustainable growth *e.g.* agriculture, mining, industry and energy should be the core concern of economic policy, and hence properly harnessed for the countrys development. The nation's technological base should be re-activated; hence the need to earnestly complete the Ajaokuta steel rolling mill, the Aladja steel plant and the Katsina, Jos and Osogbo steel rolling mills. In concert with other African states and using the instrumentality of the New Partnership for African Development (NEPAD), Nigeria should work assiduously to ensure the collaborative involvement of all stakeholders in African development through the promotion of regional economic blocs. Finally, it must be stressed that given the nature and character of the Nigerian state with its inherent weak domestic base, globalization, certainly has its adverse implications on the nation's economy. However, in order to minimize the negative effects and harness the whatever benefits of the current growing and overwhelming globalization trend; there is the necessity for a complete restructuring and transformation of the Nigerian political economy.

13
History of Nigeria Trade

13.1. History of Nigeria Trade

From the mid-1970s onwards, Nigeria's main trade policy instruments shifted markedly away from tariffs to quantitative import restrictions, particularly import prohibition and import licensing. As a reflection of this shift, Nigeria's customs legislation established an import prohibition list for trade items and an absolute import prohibition list for non-trade items. While the trade list covers the full range of agricultural and manufactured products, the non-trade list relates to goods and services that are considered to be harmful to human, animal and plant health, as well as public morals. Typical examples of products which feature on this second list include weapons, obscene articles, airmail, photographic printing paper, base or counterfeit coins and second-hand clothing. Furthermore, the customs legislation empowers the government to modify these lists at its discretion, by adding or subtracting items through customs and excise notices and government announcements.

Based on this legislation, the government placed seventy-six broad groups of import items on the import prohibition lists in 1978. The number of items placed under import prohibition increased further, particularly during 1982-5. Hence, at the beginning of 1986, roughly 40% of agricultural and industrial products, in terms of tariff lines, were covered by import prohibitions. This sharp increase in the coverage of import prohibitions abated somewhat during the second half of the 1980s; by 1989, import prohibition covered about 29% of agricultural products and 20% of industrial products measured, again, in terms of tariff lines Nigeria exports primarily petroleum and other raw materials such as cocoa, rubber, palm kernels, organic oils, and fats. It imports secondary products such as chemicals, machinery, transport equipment, manufactured goods, food, and animals. The dependence on oil and a few other commodities for export caused Nigeria to become especially vulnerable to world oil price fluctuations. During the colonial years, Britain was Nigeria's leading trading partner. After independence, Nigeria diversified its trading partners. It now trades worldwide with about 100 countries. The United States replaced Britain as the primary trading partner in the 1970s. However, Britain remains Nigeria's leading vendor, selling the former colony more than 14 per cent of its imports in the 1990s. Other major trading partners are Germany, France, the Netherlands, Canada, Japan, Italy, and Spain. Nigeria's meager trade with Eastern Europe and the former Soviet Union declined even further after the collapse of Euro- Communism and the breakup of the Soviet Union in the early 1990s. Nigeria's trade with sister African countries–mainly with other West African members of the Economic Community of West Africa (ECOWAS,

Trade (expressed in billions of US$): Nigeria

Year	*Exports*	*Imports*
1975	7.845	6.041
1980	25.968	16.660
1985	12.548	8.877
1990	13.670	5.627
1995	34.179	34.488
1998	37.029	43.798

Source: International Monetary Fund. *International Financial Statistics Yearbook* 1999.

Prior to 1966, Nigeria had a persistent **trade deficit.** The rapid growth of petroleum as an export commodity reversed the trend between 1966 and 1977. Sluggish international demand for Nigerian crude oil renewed the trade deficit from 1978 to 1983. Severe import restrictions and an economic structural adjustment program (ESAP) adopted to address the economic breakdown brought about trade surpluses from 1984 to 1986, and again in 1990. Monies sent home by Nigerian residents overseas helped to cushion the drastic effects of the deficit and the ESAP-induced decreased government spending on the population. Although particular items moved in and out of the import prohibition lists over the next ten years, the general trend in reduction in the number of items whose importation was prohibited was broadly sustained. Hence, by 1998, only 127 (out of 5, 147 tariff lines, or 2.5%) remained on the import prohibition list for trade. But with effect from late 2001 and continuing until early 2004, another upsurge in the number of items placed under import prohibition has occurred. In particular, the number of broad product groups under import ban rose from twenty-seven in February 2003 to thirty-five in January 2004.

In terms of sectoral coverage, import prohibition has focused on such agricultural products as fruit, vegetables, grains, meat and fish, as well as manufactured products including rubber, wood and cork, textiles and chemicals. In 1989, for example, close to 96% of the tariff lines for textiles and clothing were subjected to an import prohibition regime, with similar coverage ratio for several other sectors being as follows: furniture (93%), wood and wood products (45%), rubber (5%) and chemicals (1%) (GATT 1991). During 1982-5, the import prohibition coverage ratio for food, beverages and tobacco was over 50%. The pervasive use of import prohibition as an instrument of trade policy in Nigeria derives from a long-standing import policy regime which was designed to promote industry, employment and balance-of-payments objectives in the context of an import substitution-industrialization strategy (Oyejide, 1975). Key elements of this regime include protecting existing domestic industries and reducing the country's perceived dependence on imports, while at the same time ensuring the availability of raw materials and capital goods which cannot be obtained from domestic sources. With specific reference to the agricultural sector, trade policy has generally been aimed at discouraging importation of all food and raw materials that the county is deemed to have the resources to produce. In the case of the manufacturing sector, a major goal has been to increase the local content of Nigerian industrial output through enhanced use of local raw materials. The achievement of this goal is promoted by the government through various measures and incentives, including import prohibition.

Sectoral coverage of import prohibition has obviously varied over time. But it has been determined largely by the general policy that imports of certain products could be prohibited either if they are judged to be 'not essential' or when they compete with domestically produced goods that are available in adequate quantities. The various motivations for using import prohibition have, however, not been fully reflected in the justifications periodically offered by the government when import prohibition notices are issued. For instance in April 1982, when a wide range of products was placed under import prohibition, the Nigerian government notified the General Agreement on Tariffs and Trade (GATT) of the measures taken with the claim that the measures had been necessitated by unfavorable external circumstances, including a deterioration in the terms of trade and sharp declines in the country's oil revenue and foreign exchange reserves. But import prohibition was periodically used for other purposes. The almost permanent ban on the importation of textile and clothing products since the late 1970s can be explained primarily in terms of protecting local industries; while import prohibition applying to such items as gypsum, kaolin, bentonites and barytes reflects attempts to promote local sourcing of raw materials for manufacturing in Nigeria. Thus when in March 1998 Nigeria notified the WTO Committee on Safeguards that the import prohibitions on wheat flour, sorghum, millet, gypsum and kaolin were imposed for safeguard reasons, there was credible reason to question the claim. The pervasive use of import prohibition in Nigeria has another, perhaps equally important, reason: it is administratively easier. In Nigeria's responses to the questions raised on this matter during discussions at various GATT and WTO fora, it has been argued that import prohibitions are easier to monitor than price-based measures, since the presence of the banned products on local markets is, in principle, sufficient for enforcement.

13.2. Local and External Players and their Roles

It is well established that any trade policy change is likely to generate both winners and losers, even if the overall net impact is positive. Economic agents and other stakeholders involved with, or affected by, any trade policy change can be classified broadly into two groups; they are either local (domestic) or external (foreign). Each of these broad groups can be broken down into more specific interest groups, such as producers, importers, exporters, traders, workers and consumers. With specific reference to the import prohibition policy in Nigeria, local stakeholders range from the policy-making and enforcement agencies through producers of the banned imports and their workers to the importers and consumers of banned products.

On the external front, direct stakeholders include countries whose export products are denied market access in Nigeria as a result of the import prohibition policy, as well as regional and multilateral institutions (such as the Economic Community of West African States (ECOWAS), the IMF, the World Bank and the WTO) which exercise policy surveillance mandates in this area. In capturing the perceptions and views of local and external players directly involved in and affected by Nigeria's import prohibition policy, primary reliance has had to be on published official documents and the print media. Given the period over which the policy has been tracked, interviews would have focused too narrowly on more recent events and be biased by the possibility of perception, revision and rationalization. In addition, interviews may not yield useful results, as officials of government and quasi-government agencies in Nigeria are often unwilling to go on record with views that may be critical of current public policy posture.

13.2.1. Local Players

Among local players, the government and its agencies have played a prominent role in the initiation and sustenance of the policy of import prohibition. Much of the local opposition to import bans has generally been voiced by importers and traders. The consumers who ultimately bear the burden of the resulting higher prices and poor quality and limited variety of locally produced alternatives have remained largely silent, probably because they have not been organized. By contrast, the producers of import-competing goods and their employee unions have generally supported import bans.

On the side of government, policy statements and policy actions with respect to import prohibition have not always been synchronized and consistent. In particular, the apparent and often repeated decision to move away from the use of quantitative trade restriction measures has not been implemented in practice. In the 1970s, of course, policy statements and actions reflected the same restrictive trade policy posture. Thus, Roland Adeleye, Federal Commissioner for Industries, correctly reflected both policy and action when he stated in 1977 that 'the federal government will not import into Nigeria anything that is produced in adequate quantity by Nigerian industrialists' Divergence between policy statement and policy action has been particularly strong since the late 1980s. For instance, in a policy statement submitted to the WTO by the Nigerian government in 1998, it was indicated that 'the list of items removed from import prohibition continues to lengthen' (para. 14), and that 'necessary steps are being taken by government to comprehensively eliminate all existing items on the import prohibition list as soon as possible' (para. 17). Along the same lines, President Olusegun Obasanjo's foreword to the *Trade*

Policy of Nigeria (Federal Ministry of Commerce 2002) envisages a 'dynamic trade reorientation which will signify a clear departure from past regimes of controls and intervention'. Furthermore, this document affirmed that 'government shall, within the limits of its rights and international obligations and agreements, strive to eliminate quantitative trade measures' (p. 10). In practice, however, import bans actually increased in scope and coverage until 2004.

This policy had several negative effects as far as the importers of and traders in a wide range of products on the import prohibition list were concerned. Thus, the Refrigerator and Air Conditioner Dealers Association (RADA), in its statement, 'condemned the ban on importation of these products because it would lead to substantial losses of income and jobs, and further aggravate the unemployment situation' in Nigeria (*Guardian* October 2001). Similarly, the Motor Dealers Association of Nigeria (MODAN) issued a statement which argued that 'the ban on used vehicles would destroy four million jobs' (*Guardian* December 2001). In the same vein, the Embroidery Lace Dealers Association of Nigeria (ELDAN) claimed that 'an immediate enforcement of the ban on importation of textiles would inflict colossal financial loss on textile imports and eliminate three million jobs' (*Guardian* March 2004). The domestic producers of banned imports and the workers' unions associated with them not only generally lobby government to impose and maintain its import prohibition policy but also articulate its advantages. Thus, in its reaction, the National Association of Cottage Industrialists of Nigeria (NACIN) urged the government to 'ensure strict implementation of the ban on imported products as a means of guaranteeing the survival of small and medium-scale enterprises and to create employment for the nation's teeming graduates' (*Guardian* February 2004). In the statements of the Manufacturers Association of Nigeria (MAN), this point of view was pushed further by 'proposing a minimum lifespan of five years for the current import restrictions policy as a means of ensuring that it achieves the desired results' (*Guardian* March 2004). Finally, the National Union of Textile, Garment and Tailoring Workers of Nigeria (NUTGWN) 'considered the textile ban as the best development in the textile industry in recent times because of its beneficial impact on local output and employment' (*Guardian* April 2004).

13.2.2. External Players

The key external stakeholders that have played active roles in the discussions surrounding Nigeria's import prohibition policy include several countries whose exports have been directly affected and a number of multilateral organizations. Between 1980 and 1991, at least three countries lodged formal complaints against Nigeria with respect to import prohibitions: Norway submitted a complaint on Nigeria's import ban on stockfish, Côte d'Ivoire on the import ban on textiles, and the United States on the import ban on wheat and rice. While both Norway and the United States cited violation of GATT rules in their complaints, Côte d'Ivoire's case rested on a violation of the ECOWAS treaty. These complaints were settled through bilateral negotiation and consultation. More recently, both the European Union (EU) and Benin have raised issues with Nigeria with respect to its import prohibitions. Speaking on behalf of the EU, Claude Maerten, an official of the EU Trade Directorate-General, argued that 'Nigeria's import ban was not compatible with, and indeed forbidden by, WTO rules' (*Guardian* July 2003). In the case of Benin, the country's ambassador, Benoit Adekambi, issued a statement in which he claimed that 'Nigeria's import bans, particularly on textile products, had dealt a severe blow to the economy of the Republic of Benin' and 'constituted a violation of the Memorandum of Understanding between the two counties regarding continuous trade liberalization'.

The three multilateral organizations which have expressed views on Nigerian import prohibition policy can be classified in two categories. In one of these are the World Bank and the IMF, while the other consists of the WTO. Both the World Bank and the IMF have only an advisory role with respect to trade and other policy matters in Nigeria. It is in this context that both organizations encouraged the liberalization of Nigeria's policy regimes. The resurgence of import prohibition represents a sharp reversal of the trade liberalization programme initiated in the mid-1980s with the support of the IMF and the World Bank. Both organizations provided technical and policy advice and the World Bank also provided support through its lending programme. In particular, the World Bank made two quick loans to promote trade liberalization: the Trade Policy and Export Development Loan of US$450 million in 1987, and the Trade and Investment Policy Loan of US$500 million in 1989 (World Bank 1994). Both had as part of their objectives the reduction and eventual elimination of import prohibition. Although these loans were fully disbursed, subsequent evaluations concluded that the import regime reform encountered strong opposition both within and outside the government and that Nigeria had, in general, displayed a poor implementation record and commitment to import liberalization (Castillo, 1993). More specifically, the World Bank, in its assessment of Nigeria's trade policy reform, sketched the following sequence of events (1994: 11-12):

As it transpired, the SAP's attempts to achieve transparency and stability in the incentive system were overtaken by events. The list of banned imports was once again extended so that by 1991, about 20% of industrial imports and 30% of agricultural imports were affected. About 1, 000 of the harmonized system of 5, 000 six-digit imports, furthermore, remained subject to conditional import prohibitions, which could be invoked on the basis of balance-of-payments considerations. Nigeria's import prohibitions and their justifications in terms of balance-of-payments problems have triggered many discussions in GATT and then the WTO since the early 1980s. Nigeria first invoked GATT Article XVIII:B on import restrictions for balance-of-payments reasons in 1982. This led to the first consultation with the GATT Committee on Balance of Payments Restrictions in April 1984. In notifying its import restriction measures to GATT in April 1982, Nigeria emphasized their temporary nature. The Committee recognized the serious balance-of-payments problems faced by Nigeria in calling for the introduction of the restrictive measures, but encouraged it to pursue more appropriate economic stabilization policies. Follow-up simplified consultations with the Committee were held in October 1986, October 1988 and March 1991. At these meetings it was noted that in spite of the removal of some of the restrictive measures, several import bans introduced for balance-of-payments reasons remained in force. Finally, in 1996, the WTO Committee on Balance of Payments Restrictions decided that Nigeria's import prohibitions could not be justified under the balance-of-payment rules of GATT 1994. The consultations in February 1998 discussed Nigeria's import bans on maize, vegetable oils, barytes and bentonites, as well as plastic articles. Although the consultations closed without agreed conclusions regarding Nigeria's phase-out programme for the restrictions, members of the Committee stressed, once again, that the import bans were not consistent with WTO rules. Thus Nigeria's use of import prohibitions as a trade policy instrument has been a source of friction with its trading partners; this practice has also been repeatedly condemned for its inconsistency with GATT and WTO rules.

13.3. Challenges Faces and the Outcome

In using import prohibition as a major trade policy instrument, Nigeria has hoped that its balance-of-payments problems would be alleviated, and that the protection offered would induce increased output and employment of the domestic industry. Against these postulated positive outcomes must be set several possible negative consequences of import prohibition, including raising the domestic prices of import-banned products, disrupting other sectors which use the prohibited imports as raw materials, depriving government of tariff revenue and creating vested interests among domestic producers of prohibited products and among smugglers. Nigeria's balance-of-payments situation is determined primarily by developments in the world oil market; hence it has not been amenable to changes induced by import prohibitions. In any case, it seems clear that protection of domestic producers is the real force behind the use of this policy instrument. But there is little evidence that it has produced the desired result here either. For instance, a recent study of the textile sector – the single most important target of import prohibition policy of the past twenty years – shows that both its output and employment have stagnated or declined (Oyejide *et al.,* 2003). In addition, a survey of manufacturing-sector performance conducted by the Manufacturers' Association of Nigeria (1989) does not support the view that the level of capacity utilization was positively related to the degree of local sourcing of raw materials – one of the major channels through which import prohibition was expected to promote increased output and employment.

There appears to be recognition both within government and among producers that the import prohibition policy is rendered virtually impotent by large-scale smuggling and that this has continued in spite of stiff penalties imposed on those involved with the importation, transportation, storage, display or sale of prohibited items. This recognition has not, however, led to the abandonment of the policy. Rather, pressure has mounted to enhance its stricter implementation. According to Moses Gbadebo, president of the National Union of Chemical, Footwear, Rubber, Leather and Non-Metallic Product Employees, 'it is not enough to say we have banned items, the government's words must be backed with action the law should be properly enforced' (*Guardian* January 2004). Similarly, according to Nasir Lawal, president of the National Union of Textile, Garment and Tailoring Workers, 'if you place a hundred bans and your borders still remain porous, it will only help in increasing the income of smugglers and their agents' This concern is, apparently, fully shared by Olusegun Obasanjo, the Nigerian past president, who recently accused the Nigerian Customs Service of 'making nonsense of government's import prohibition policy'. In frustration, he was reported to have said 'we just have to beg them. I think other than begging, I don't know what else I can do. If it is possible to run a nation without customs, I will do it'.

The apparent reluctance to abandon prohibition policy is reflected in Nigeria's responses to issues raised during various GATT and WTO discussions on this matter. For instance, during the consultations in 1984, Nigeria pledged to eliminate its import prohibitions quickly. Over time, many of these import bans were indeed lifted, but the policy itself was not abandoned. Following the 1996 decision by the WTO Committee on Balance of Payments Restrictions that Nigeria's import prohibitions could not be justified under WTO rules, Nigeria offered to eliminate all such measures by early 1997 – only to begin to notify additional ones in 1998. A further proposal was made by Nigeria to phase out all remaining import prohibitions by 2005 under an eight-year elimination programme (WTO, 1998). This proposal argued that this period was necessary to allow time for the ongoing customs and port reforms to take root so as to ensure the effective administration of the resulting price-based measures and to allow the economy to consolidate the recent gains in the area of inflation, external reserves, interest and exchange rates. The upsurge in the use of import prohibitions during 2001-4 raises considerable doubt with respect to Nigeria's commitment to its own import prohibition phase-out programme.

13.4. Lessons

Several lessons can be drawn from Nigeria's import prohibition policy experience. Perhaps the most general of these is that the coherent and consistent pursuit of good trade policy requires not only a robust and appropriate domestic institutional framework and process for trade policy-making but also a supportive and institutionalized multilateral arrangement for trade policy surveillance. Weaknesses in both of these may be responsible, in varying degrees, for the persistence of Nigeria's import prohibition policy. Nigeria's internal trade policy surveillance mechanism consists largely of the domestic framework and trade policy-making process, both articulated at length in the *Trade Policy of Nigeria* (Federal Ministry of Commerce, 2002). However, the country's actual trade policy-making deviates quite substantially from what this document stipulates. These deviations largely explain the lack of coherence between policy statements and policy actions; this probably also derives from the absence of local ownership of the trade liberalization policy which appears to have been induced by the World Bank-IMF supported structural adjustment programme.

Within the internal trade policy surveillance mechanism in Nigeria, vested interests built around the use of quantitative import restrictions have acquired tremendous powers, and no effective counterweights have evolved over time, perhaps because consumer groups and other civil society organizations are not sufficiently well organized. Furthermore, the mechanism lacks effective feedback systems in the articulation, implementation and evaluation of trade policy. In particular, trade policy initiatives are not routinely subjected to cost-benefit analysis; and no explicit system of monitoring and evaluation is used for modifying or changing trade policies which do not produce desired results efficiently. Nigeria's membership of the WTO provides it, in principle, with a strong external trade policy surveillance mechanism. But the role of the WTO as an 'agent of restraint' in favour of good trade policy is feasible only to the extent that two important conditions are met. First, the government whose behaviour is to be 'restrained' must be committed to good trade policy and thus be willing to tie its own hand and use an external treaty obligation to strengthen its hand against local vested interests. Second, the external agent must have adequate sanctions which it is able and willing to use to punish deviations from the pursuit of good trade policy. Neither of these conditions appears to have been effective in dissuading Nigeria from the continued use of its import prohibition policy. The basic problems inherent in the framework and process of trade policy-making in Nigeria and its surveillance are not unknown. Some of these are eloquently summarized in a recent statement by Olusola Faleye, president of the Lagos Chamber of Commerce and Industry (*Guardian* April 2004):

We have no query with the principles of protection, but with the process. We are concerned that the recent import prohibition policy was not preceded by sufficient consultation. Import prohibition is a major trade policy decision which requires wide-ranging consultations, capacity surveys and the advice of trade and economic development experts before being pronounced as policy. Sufficient account was not taken of the local capacity vis-á-vis local demand and issues of policy transitions and implications for existing treaties to which Nigeria is signatory. Any sustained effort to eliminate import prohibitions in Nigeria is unlikely to be permanently successful if it does not first address these underlying problems.

13.4.1. Barriers of Trade

He subject of trade barriers has been a recurring issue in the country for years. Thus, a forum that emphasised the need for African countries, West African countries in particular, to trade with each other was organised Thursday by the Lagos Chamber of Commerce and Industry (LCCI). The event, tagged: 'Stakeholders' Forum on Breaking Trade Barriers in West African sub-region,'drew participants from across the country. Joseph Annan, deputy minister of trade and industry of Ghana, gave the keynote address, while Lewis Olu Ogunojemite, comptroller of Customs, presented a paper at the forum, and recalled Hillary Clinton, US secretary of state, calling on African countries to increase trade with each other, in other to realise the enormous potential that exists within the continent. The customs boss argued that although ECOWAS have made strides towards free trade, "full market integration remains an ambition. The efforts to eliminate tariffs have begun and all members, except Liberia, have eliminated tariffs on unprocessed products. Finished products also only attract the Export Trade Liberalisation Scheme (ETLS) tax; revitalisation of trade liberalisation has begun with the fast track initiative between Ghana and Nigeria; efforts are on going to remove non-tariff barriers, however, progress on removal of these barriers is harder to assess, as data on such barriers are inadequate and by their very nature, they are not directly measurable."

For him, there are other underlying structural constraints such as deep-seated distrust among member states, which fundamentally limits the depth and progress in regional integration, there are systemic problems that hamper the development of national economies and invariably impede the regional integration, there is the reluctance to adhere to integration programmes due to concerns over losses and uneven gains, insufficient analytical and technical support also limit implementation of some integration instruments, for instance, the trade liberalisation scheme. In terms of progress made in advancing the goals of regional integration, Ogunojemite said the results are mixed with a general sense that there remain substantial gaps in achievements." Some encouraging advances have been made in the area of information and communication technology (ICT), due to global advances in this sector, as well as the increased role of the private sector in enhancing service provision. However, in comparison to other parts of the world, the sub-region is still a long way from realising universal access to ICT," he argued.

Quoting World Bank (2003), Hester Hopkins, general manager, operations, COTECTNA, said trade can be a powerful force for growth and poverty reduction. Countries that have increased the share of trade in their GDP have grown faster and reduced poverty more rapidly. She stressed the need for trade facilitation in the sub-region to reduce the complexity and cost of the trade transaction process, and ensure that all these activities take place in an efficient, transparent and predictable manner. Those she listed as important players in the process of trade facilitation include: government and government agencies, transporters/logistic providers, financiers/insurers, and traders. For Hopkins, trade facilitation reduces transaction costs, allows for faster delivery/clearance, predictable trade rules, increased foreign investments, reduced corruption, increased revenue and increased compliance. For her too, trade facilitation principles include harmonisation of applicable laws and regulations, simplification, that is the use of ICT to exchange information efficiently, standardisation which has to do with administrative and commercial formalities, procedures and documents; transparency, that ensures that information, requirements are processed for crossing borders, clear, specific and easily accessible for all involved.

Citing the SWEPRO/National Board of Trade, Sweden, 2002, she argued that trade facilitation results in direct benefits both to governments and to the business community in terms of increased economic efficiency, better security, faster delivery of goods, and reduced costs. Hopkins argued that the World Bank observes that the top 10 trade facilitation economies, which include Singapore, Hong Kong, China and the Republic of Korea, share four characteristics as follows:

- Introduced paper-free electronic data interchange systems and pre-arrival submission of customs declarations.
- Make use of a risk management inspection system, where less than 10 per cent of containers are physically inspected.
- Have systems allowing authorised persons, who meet prescribed criteria to benefit from simplified procedures.
- Have excellent port infrastructure.

Roberts Orya, managing director/CEO, Nigeria Import Export Bank, also reeled our data on the economy of the sub-region and place of Nigeria in the economy."

According to IMF/World Bank data, the ECOWAS region attained a total GDP of $565 billion and a population of 289 million in 2010. Of this figure, Nigeria contributed over 60 per cent, with a GDP of $374 billion and a population of 150 million. The Economic Community of West African States (ECOWAS) Treaty, which is a regional grouping of 15 member countries, is expected to eliminate all tariffs, restrictions and quotas towards facilitating free movement of people, goods and services, thereby promoting social and economic harmonisation, deepening trade and encouraging large-scale investment. The above mentioned goals are expected to achieve the attainment of a free trade area, a unified custom union/establishment of a common external tariff, creation of a common market, and the creation of an economic union (harmonisation of economic, agricultural, industrial and monetary policies, etc. Orya further said: "Intra-regional trade remains a relatively small part of the ECOWAS economic activity, accounting for about 10 per cent of the total trade of the member states. In 2008, the total ECOWAS intra-regional trade was valued at $6.9 billion, while total ECOWAS trade with the world was valued at $64.4 billion."Historically, the value of ECOWAS intra-regional trade has over the years been increasing from $3.2 billion in 2002 to peak at $9.9 billion in 2006 before it nose-dived to $6.6 billion and $6.9 billion in 2007 and 2008 respectively." In tonnage, the volume had similarly increased from 4.2 million tons in 1996 to 13.4 million tons in 2008."

For Orya, "With regards to Nigeria's level of ECOWAS intra-regional trade, the aggregate non-oil exports stood at $189.96 million in 2009 declining from $228.81 in 2008, and representing 9.7 per cent of Nigeria's non oil export in 2009 account to the 2009 CBN Annual Account." The low level of ECOWAS intra-regional trade, according to Orya, are largely traced to some tariff and non-tariff barriers, some of which include lack of appropriate/adequate infrastructure at the region that is characterised by lack of poor rail systems, roads, energy etc. - this is more compelling when viewed against the empirical evidence that cargo volumes have leaped about 300 per cent from 4.7 million tons in 1998 to 13.4 million tons in 2008 without corresponding increase/improvement on basic logistical infrastructure; barriers to movement of goods and services, due to multiple check points/border posts with attendant consequences of time and cost overruns and eventual uncompetitive pricing of goods at market destination. Femi Deru, president, LCCI, said trade within the ECOWAS sub-region has become more imperative today than ever before. "Global competitiveness is a major factor in achieving efficiency, which is in itself dependent on market size and the ability to enjoy economies of scale," he said. For him, with a robust market of about 300 million people, significant benefits of economies of scale would be enjoyed by West African firms in the event of full market integration.

Deru argued there is a strong relationship between the development of intra regional trade and the promotion of economic integration in the sub-region. "Integration is in fact the main vehicle for the boosting of trade within the sub-region, with integration, African economies would be stronger and their capacity to cope with the challenges of globalisation would be enhanced. Africa would also be better positioned to respond to the challenges of peace and security in the continent," he said. For him, some progress has been made under the current integration initiatives in Africa, but on the whole, the performance in economic integration has not been quite satisfactory. Joseph Annan, who delivered the keynote address, addressed the issue of constraints to West African trade listing high transport cost and poor infrastructure and border measures and regulatory environment. He argued it is well known that transportation costs are extremely high between our countries. He said road freight tariffs in sub-Saharan Africa are in the range of $0.04-0.14 per ton per kilometre compared to $0.01-0.04 per ton per kilometre in other developing regions. "Trade to landlocked countries in particular is especially costly, with import costs three to five times the global average. It takes 116 days to move a container from Bangui, Central African Republic to the nearest port," he said. For him, time-consuming and costly border measures and unharmonised and cumbersome regulatory systems are even greater barriers to trade than poor physical infrastructure.

Annan, the implementation of existing agreements is the most important step; negotiating new agreements could even be counterproductive. He said since the implementation of regional agreements can only advance at the pace of the slowest moving country, pressure from peers and business leaders is needed to speed this process. He advised that there is need to enforce existing trade protocols especially ETLS and free movement of persons. For him, at the national level, countries need to think carefully about how to promote West African trade, and then act to integrate these strategies into their development plans.

13.5. Trade Policy and the Poverty Reduction in Nigeria

Evaluating trade policy (particularly the 2002 trade policy document in Nigeria) and the extent to which it facilitates or hinders poverty reduction in Nigeria. It examines the state of poverty and the poor in the country and proceeds to evaluate the provisions of the current trade policy document. Based on the intention to provide inputs into the ongoing review of the current trade policy, it also evaluates its consistency with the NEEDS and other domestic development policy documents. Since the last twenty years, economic policy in Nigeria can be characterized by trade liberalization and regional integration which is defined by the radical reducing or removal of trade barriers. The World Trade Organization (WTO) the IMF and the especially the World Bank (WB) have obtained considerable powers to sway policies in countries towards this path. As a part of the global structural adjustment programme, it is assumed and argued that trade liberalization improves the welfare of consumers and trims down poverty. The assertion was two-fold and simple. First, it is argued that liberalization offers wider room for choice from an array of quality goods and cheaper imports also find more lucrative markets in which their products can be sold. A second argument is that, the production of goods in which a country has comparative advantage expands, while the sectors without comparative disadvantage minimize. This is believed to lead to an overall rise in real GDP since there would be reallocation of the productive factors from less efficient sectors to more efficient sectors. 'It is against this setting that many developing countries liberalized their imports, reduced average tariffs and dismantled significant number of Non-Tariff Barriers.

Nevertheless the degree of realization of the anticipated rewards of liberalization over the years of agonizing alteration has been the topic of intense debate in many empirical literatures. Some scholars believe these rewards have come, others think it is on the way, while yet another group believe it neither has nor will ever come. Those who give emphasis to trade liberalization also stress the necessity, as a minimum, for an established exchange rate system, cautious monetary and fiscal policies, good governance and a corruption-free Administration of economic policies for trade liberalization to be effective in the long-run. Those against it point to time discrepancy of policy makers, doubts about losses and gains of liberalization processes, increasing barriers to trade by developed countries and unfair competition among other things. The debate rages on and thus far is still far from being decisive on either side. Notwithstanding the fact that the debate is yet to be concluded, policymakers in developing countries are practically being pushed to liberalize. Meanwhile, the gains and costs from the revolutions spun by globalization and integration of world trade are unevenly dispersed and it is not apparent that those who get the best benefits also have the highest cost. Rather to the contrary, evidence show that economies of developing countries often carry the burden of increased openness. The existing Nigerian trade policy document is one of the numerous drawn with strong focus on the global economy, displaying outward orientation in all aspects. Questions thus arise as to the implications of this for poverty reduction which is vital for comprehensive growth and development. It consequently becomes imperative to realize the degree to which trade policy can be used as a successful tool for poverty alleviation. Evaluating the effectiveness of trade policy in promoting livelihood of the poor in Nigeria by evaluating:

- ☆ The degree of strengthening or alleviation of the existing marginalization of the poor by trade policy.
- ☆ The instruments and provisions within trade to release the poor from the poverty plague.
- ☆ How trade can smooth the progress of growth with equity and a better allocation of national wealth for instance, by adding value to Nigeria's primary products such as oil and agriculture products.
- ☆ How trade can facilitate the development or attainment of technology and enhancement of local capacities.
- ☆ How trade can facilitate investment in human capacity including education, health, food, housing, etc

The following sections will attempt to show the relationship between international trade and poverty issues, examine the views of operators in the real sector on the impact and trade Policy, Gender and the Livelihoods of the Poor in Nigeria possible directions of trade policy. This goes to contribute to the debate on the impact of international trade on poverty reduction using the experience of

Nigeria. The rest of this paper is organized as follows: Chapter two presents a global view of trade, economic growth and poverty focusing on theoretical and empirical links. Chapter three relates the relationship of trade, macroeconomic policy and poverty reduction to Nigeria's experience and history. Chapter four discusses the present trade policies of Nigeria and finally, chapter five summarizes the key emerging issues and recommendations of this paper.

14

Trade, Economic Growth and Poverty: Theoretical and Empirical Links

14.1. Trade, Economic Growth and Poverty: Theoretical and Empirical Links

The basic argument is that international trade can play a powerful role in reducing poverty in the least developed countries as well as in other developing countries. But the national and international policies, which can facilitate this, must be rooted in a development driven approach to trade rather than a trade-driven approach to development. An exclusive focus on trade, which assumes that poverty is reduced through trade rather than through development, is likely to prove counter-productive. Rather, it is necessary to look at the relationship between trade and poverty from a development perspective. These was the opening thoughts of UNCTAD Least Developed Countries' Report 2004. Trade and Development", Trade and Poverty Reduction " are all sub themes among many others of a major theoretical and empirical question that has tugged at the heart of economic debate since the publication of "Wealth of Nations" by Adam Smith. However, the clichés have multiplied with the push for reforms and liberalization of trade in various developing and transitional economies over the last thirty years. What therefore are the issues, what are the arguments and what are the alternatives in approaching international trade? Are there really any link between trade on the one hand and growth improvement in welfare and poverty reduction on the other? Are there really any empirical backings to the theoretical noise? Should UNCTAD and other groups, who hold the view that there is some sort of relationship between trade and poverty, be taken seriously? And why is the substance of the case made for or against trade liberalization and internationalization of trade as weapon for poverty reduction?

Certainly the connection between trade and poverty reducing growth is neither linear nor habitual. For developing Countries, many which form key targets for poverty reduction strategies and policies, UNCTAD 2004 attributes this to the technical and financial troubles related with developing a variety of new capabilities, institutions and services to adjust to the challenges brought about by changing international trade circumstances, rules and commitments. Additionally, exports are barely able to expand import capacity in many Least Developed Countries due to the structural shortages in their export basket which is typically mainly mono-product (agricultural) with high instability in international demand and market prices. Under such situations, aid and private remittance become a key source of financing imports. Thus, the link between trade expansion and poverty reduction can be broken if increased trade is associated with decreases in aid or increase in debt service responsibilities. It has been preached that while national economies of most Least Developed Countries are greatly integrated with the rest of the world, the lives and livelihoods of the people are not so linked with the international economy. Compared to a number of other arguments both for and against trade liberalization as an instrument of poverty reduction, the UNCTAD stance could be considered quite benevolent. Some vigorous arguments have been raised to prove that international trade liberalization is the progenitor of poverty and that particularly for Least Developed Countries, there is no worse prevention to growth since the 1980s than the IMF/World Bank-imposed trade liberalization and associated reforms. Chang (2000) forcefully argues that the 'bulk of today's developed world did so under protectionist policies but would not allow present day least developed countries to grow with the same hedges they enjoyed'. On the other hand, argued by the Stolper-Samuelson theorem4 and certainly by the whole range of neoclassical endogenous and associated theories, there have been anxious efforts by the pro-liberalization group to show that trade liberalization over the twenty years or more have led to higher growths in countries that 'behaved' while the non-compliers suffered terribly. The facts and figures range from the factual and linkable to the outright peculiar. This book will briefly look at the arguments and facts on both sides of the argument. The Australian International Development Agency website www.ausaid.gov.au clearly announced 'No country that has closed itself off from world trade and investment has grown for long or reduced poverty'. It subsequently rolled out a lot of figures to support the assertion that free trade boosts the income of the poor and reduces global poverty as follows:

- ✰ A study of 73 developing countries, including China, India, Brazil, Mexico, Malaysia and Philippines, found that those countries increasing their openness to trade have grown on average faster than those that did not, and faster than most. The Stolper-Samuelson (1949) theorem predicts that trade liberalization should lead to an increase in the remuneration of the relatively abundant factor of production in low and middle income countries(unskilled labour) with respect to the scarce factor (capital or skilled labour) rich countries. Annual incomes in these countries grew an average of 5 per cent in the 1990s –significantly higher than previous decades. Notably, there countries reduced tariffs by an average of 34 per cent point since 1980, increasing their trade exposure. By contrast, economies that did not reduce tariffs experienced a lower average annual income growth of 1.4 per cent in the 1990s.

☆ Between 1993 and 1998 the number of people living in absolute poverty in developing countries that opened themselves to trade declined by 14 per cent. By contrast, poverty in developing countries that did not open themselves to trade rose 4 per cent between 1993 and 1998. Global poverty over the last two decade has reduced by some 200 million people, while school enrolment and life expectancy have increased.

The study divulged that at times free trade could inflict some costs as industries make alterations to contain greater international competition, but also promptly added that 'other sectors prosper as a consequence of greater access to world markets and over time they absorb workers who lose their jobs elsewhere. Even for the very poor, freer trade especially in agriculture, allows them to buy basic goods more cheaply 'In a WTO ministerial roundtable meeting in London in 2001 and cited in the United Kingdom Department for International Development 2002 (DFID) background briefing, it was announced that 'In recent decades, it is those countries which have seized the opportunity offered by more open world markets to increase exports and attract inward investment that have made the greatest strides in reducing poverty'.

14.2. Linking Trade Liberalization to Poverty

Theoretically the factors linking trade liberalization to poverty include the following:

1. Changes in the Price of Exports and Imports

By reducing prices of imported goods while at the same time keeping the prices of import substitutes low, trade liberalization would be able boost the real incomes of the poor. It increases the accessibility of products that might be very critical to the poor such as basic food, pharmaceuticals or basic health products, clothing and other products. Likewise, by getting rid of export barriers, trade liberalization can increase the prices producers receive in addition to stimulating the production of exportable; thus leading to increase in employment and income of the poor. It can also allow the importation of new technologies and processes.

2. Vulnerability to External Shocks

The poor are mainly awfully susceptible to external shocks and trade liberalization amplifies the likelihood of such shocks. Trade liberalization can make an economy more susceptible to external shocks for instance through sharp changes in the terms of trade which can turn the fate of a large proportion of the population especially when it affects such sector as agriculture. Whole generations can be thrust back into poverty and could also cause an intergenerational transfer of poverty to arise from such shocks particularly when it also leads to a drop in social spending.

3. Factor Prices, Income and Employment

It is assumed that shifts in resources cropping up from production rearrangement, based on comparative advantage, accompanying trade liberalization could have significant effect on income and employment. This occurs as the demand for traditional exports of the country increases leading to an increase in the real wage to those employed in the sector. The Stolper-Samuelson (SS) theorem envisages an increase in the real returns to the factors used on the production of a good when its relative price in the international market rises.

However, in reality, there are quite a number of labour market inflexibilities that may make this prediction unfulfilled even when there is an increase in the relative price of a product.

4. Government Revenue and Programmes for the Poor

Revenue to governments from trade taxes frequently falls when tariffs are reduced. Soludo *et al.* (2003) predicts such a drop for revenue in the event of latching on to the ECOWAS Common Economic Tariff. This has often reduced the competences of governments to invest in domestic infrastructure and offer support to domestic production hence, leading to the cutting down of social spending, which in turn negatively affects the poor. But this also does not have to be particularly so if competence of collection is enhanced, exemptions reduced or placing tariff on non-tariff products. Such development in collection competence can be the instant overcome of reduction in overall tariff rates. Besides, at the regional level, liberalization can lead to the giving up of the delivery of basic social services to non-national private sector operators which can additional have adverse effect to the poor. This can also lead to the dumping of surplus manufactured and agricultural products in developing countries by their more developed colleagues, unfavorable competition for government procurement by foreign firms at the expense of local producers. The theoretical link is that the effect of liberalization on growth and poverty in a country depends decisively on the structure of its exports and imports. Frequently high export oriented countries gain while high importers lose. This has been the case with countries in Sub-Saharan Africa including Nigeria, who have little to export and are experiencing deteriorating product prices but are obliged to liberalize their trade. These countries habitually end up with more foreign product leading to a balance of trade deficit. Following Nigeria's membership to the WTO and reduction of tariffs and other commitments, imports as a proportion of GDP rose exponentially, infuriating its balance of payments and debt issues. Imports rose from 8.18 per cent of total trade to 37.4 per cent in 2001 with an all time high of 42.3 per cent in 1999.7 The anticipated increase in non-oil exports did not happen while imports flowed. The experience of Nigeria shows a flow in imports and with the exception of oil and gas, while there is little to export. Raising the question of how would the resources to finance the imports is generated if oil was not available? Efficiency and exporting are highly related because efficient firms exports.8 In a case where the supporting infrastructure and environment for efficient production is absent and the quality of human capital is low, liberalization will attract more imports than exports and this will have unfavourable effects on economic growth. The inefficient production process and the goods that originate from them cannot compete against international standards. An import regime without the capacity to raise resources from export is unsustainable either in the short or long run – classic case of the *Dutch Disease.*

The overall country experiences from the above study show that:

- ✰ Economic integration, counting trade liberalization is not a prerequisite for growth but rather often an outcome of it. Trade Policy, Gender and the Livelihoods of the Poor in Nigeria.
- ✰ Countries have to manage their domestic institutional innovations and flexibility and not rely on international coercion to develop these if they are expected to grow.
- ✰ Poor countries with low social indicators generally export primary products (agricultural) with minute value added and these exports over and over again fail t raise skills, productivity or technological change. Wealthier countries, on the other hand, with high literacy and infrastructure are better placed to engage in production and trade that raises productivity, generating better human development and trade benefits. Still after liberalization, there are still a very small number who have managed to make a transition from one form of production and export to the other.
- ✰ Balance of payments problems created by premature and hasty liberalization worsen balance of payments problems, which contribute to debt and financial crisis.

  Further, continued indebtedness or financial crisis can hinder the ability of developing countries to trade. Debt reimbursement is known to divert resources from poor countries to rich and developed one and lowers the ability of governments to empower the poor.
- ✰ Ideally, mounting exports can encourage economic growth in a poor economy like Nigeria. It can increase the capacity utilization as external demand for exports rise: improve physical and human capital investment due to enhanced returns arising from new opportunities brought by transfer of technology or increases efficiency imposed on the domestic economy as a result of its exposure to the outside world; likelihood of export-enhanced type of industrialization in which the domestic labour-force up till now engaged in the agricultural sector will be integrated into semi-processed manufacturing activities; and improved balance of payment situation. But such export growth is the product of target strategies to improve on domestic productivity and enhance technological capacities and not of unguided liberalization.

14.3. Trade Policy Macroeconomic Performance and Poverty Reduction in Nigeria

Trade policy regimes in Nigeria have historically been more like responses to both positive and negative shocks in the world economy as they affect the country. Accordingly, the three distinct shocks Nigeria witnessed were followed by three different trade policy reactions. The first is the oil boom era immediately followed by the collapse of the oil boom and later the liberalization following the adoption of structural adjustment programme in the country. The oil boom came immediately after the Nigerian civil war and witnessed the Open General Licensing (OGL) with special import license on some items. There was a huge reduction in import duties for 'essential commodities' like food items (which prices rose during this period), raw materials and capital goods. The collapse of oil prices led to another regime of trade policy. This regime was to a great extent influenced by the difficulties associated with high indebtedness, balance of payments disequilibrium, and drastic drop in external reserves. These led to enactment of the Economic Stabilization This period was one of the *ad hoc* trade measures. About 235 items were placed on compulsory license while another 49 had their tariff raised. For other goods, compulsory advance deposit became a condition for their importation Beginning in 1988, and soon after the adoption of Structural Adjustment Programme (SAP), there were some attempts to systematize trade policymaking with the introduction of a seven year tariff schedule which followed the overall SAP framework. This regime tripled the tariff classifications from below 2000 lines to over 5000 lines, reduced tariff dispersions and weighted average tariff by about 10 basis points. Some attempts were still made to protect the domestic industries with higher tariff rates on agricultural and industrial imports that domestic substitutes. Raw materials and intermediate goods were allowed duty free while some surcharges for port development (5%), raw materials research and development council (1%), and freight rate stabilization surcharge (0.02%). Nigeria is among the 20 poorest countries on earth. Its poverty is deep, widespread, pervasion and deterioration. The whole landscape literally stinks with lack and scarcity. Between 1980 and 2001, the percentage of Nigeria's population in dare poverty has increased from a sheer 26 per cent to about 70 per cent. There are of course minor debates about the precise poverty numbers, the authencity of the methodology of the poverty survey etc, but there are no questions as to the fact that poverty has exacerbated over the last twenty years. Nigeria's per capita GDP has dropped by more than one third while population has grown exponentially around 3 per cent a year over the period. The table below shows the trends in poverty over the last two decades.

14.3.1. Poverty Trends in Nigeria (Per cent) of Population in Deep Poverty

An outstanding fact about Nigeria's kind of poverty is that it is in the midst of plenty (WB1996). Nigeria is the 6th largest producer of oil in the world and over a period of three decades has earned over US$350 billion from oil exports. Yet, it is the poorest OPEC member country. Thus, in the view of Soludo *et al.* (2003B), Nigeria has no business being poor. One thing however that is apparent is that a great deal of the earnings from oil was misspent on uncontained lifestyle by those with political power. Corruption has progressively developed in capacity and articulacy until Nigeria attained the exceptional achievement of being rated the world's second most corrupt country. Rent-seeking and other vices practically handicapped ingenuity and Nigeria has a very heavy debt overhang with consolidated domestic and foreign debts standing at about 150 per cent of Nigeria's GDP. Policy discrepancy predominantly over the last twenty years has lessened public spending and investment in the social sector to bare zero. Therefore, the educational system is rapidly losing its significance as the equalizing factor in the society and posing the threat that poverty in Nigeria can be *'dynastic' i.e.* children of the poor having the tendency of ending up poor. The poor also suffer the worst of lack from even the little available public utilities especially with the current privatization programme of the government. An additional trait of poverty in Nigeria is its diversity and extremity across the states of the federation. Nigeria is an extremely diverse country and so is the occurrence of poverty across the country. The Central Bank of Nigeria (CBN) 1998 poverty survey results, the Trade Policy, Gender and the Livelihoods of the Poor in Nigeria proportion of the population whose consumption is less than the poverty line is about 34 per cent. However, this varies from one state to another. For some states like 'Abia and Anambra States, the proportion is about 14 per cent and 16 per cent respectively, while for others like Bauchi, Gombe, Jigawa, Kebbi, Sokoto, Zamfara states, proportion of people living below the poverty line is close to 60 per cent. Poverty is relatively more concreted in the Northern parts of Nigeria. Almost all the states with poverty levels above 50 per cent are in the North and this concentration of extreme poor in the North have continues to show up all poverty surveys in the country.' The challenge posed by such broad allocation and outreach of poverty is that anti-poverty programmes, to be successful, have to be geographically targeted.

In discussing the connection between trade and poverty in Nigeria, an appropriate first point will be to differentiate between correlation and causation. Many authors simply attribute deteriorating poverty in Nigeria to trade liberalization in a causal relationship however cautiously skip the meticulous evidence needed to ascertain such causality. It has to be recognized yet again that the connection between trade and poverty is neither linear nor automatic. There have been parallel movement between liberalization of trade and deterioration of poverty, however, at this point it may be difficult to conclude that trade liberalization is the main cause of poverty. Numerous factors, plus institutional, structural and political processes have continued to interrelate to influence human development. It is a fact that job losses in the textile sector as firms were incapable of competing with cheap alternatives (mostly second hand clothing) and the fall in agricultural prices as imported food items submerge the domestic market are examples of the depressing effect of trade liberalization that deprive the people. It is not possible to attribute the epileptic energy circumstances in Nigeria, rundown and dishevelled infrastructure and insufficient attention to schools and hospitals which also led the poor to trade liberalization although there may be probable results of the larger range of structural adjustment and socio-economic restructuring.

An evaluation of Nigeria's poverty reduction programmes since independence shows a quite unsatisfactory result. From the National Development Plans to SAP and PRSP, the affirmed purpose has always been to shrink poverty. But this has always been unsuccessful due to several reasons which include: lack of appropriate Statistics for successful planning; institutional contention and divergences owing to duplication of objectives, corruption at all levels of society; no successful long-term plans for the nourishment of the programmes; etc. The general result is hung disaster in delivering essential services to the people and the increasing frequency of poverty. The most recent of these is the effort by President Obasanjo administration in setting and meeting poverty reducing targets through National Poverty Eradication Programme NAPEP.

14.4. Present Trade Policy: An Overview

The present trade policy is the Nigerian Trade Policy of 2002, divided into six sections as follows:

The introduction of the policy document presents the domestic and international context within which the policy was drafted. It emphasized the vital role of globalization and other developments in the global trade sphere in determining the policy, principally, the coming into effect of the WTO in 1995 which to a large extent currently influences international trade. It also emphasized the need for flexibility in the implementation of international best practices as a necessity for success and growth in Nigeria's trade.

Section 2 Article 2.1 of the policy document states that the policy shall "encourage production and distribution of goods and services to satisfy domestic and international markets for the purpose of achieving and accelerating economic growth and development".

The fundamental objectives of the policy include:

- Integration of the Nigerian economy into the global market through the establishment of a liberal market economy;
- Promotion and diversification of exports in both traditional and nontraditional markets;
- Progressive liberalization of the import regime to enhance competitiveness of domestic industries;
- Effective participation in trade negotiations to enhance the achievement of national economic gains in the multilateral trading system, as well as regional and bilateral agreements;
- Promotion and development of domestic trade including intra state and interstate commerce;
- Achievement of accelerated economic growth and development through trade within a network of public-private sector partnership;
- Promotion of the transfer, acquisition and adoption of appropriate and sustainable technologies to ensure competitive export oriented industries;

- Attraction of foreign capital inflow into export oriented production through special incentive packages and domestic support;
- Promotion of regional integration and cooperation; and
- Effective utilization of information and communication technology in electronic commerce.

Within the scope of the trade policy, the role of the government was redefined as a facilitator. Government is expected to extricate from business and let the private sector take over the engine room of the economy and ensure economic growth. The macroeconomic structure is expected to maintain the deregulation of interest and exchange rates, guarantee fiscal discipline, decrease the debt burden and liberalize the trade system. The government is also expected to run an open rule-based, market oriented economy that would give full fortification for private rights and boost domestic production. The policy proposes to affect trade in such a way as to reduce the reliance on oil by encouraging the production and export of non-oil commodities. The policy targets economic growth rate of 7 per cent within the medium term fuelled by non-oil exports, which is expected to grow by 25% The policy foresees an increase in oil reserve to 40 billion barrels and a rise in productive capacity to 4 million barrel per day. It also try to add value to crude oil through processing before exporting and fully link up the oil sector to other part of local economy. Essentially this seeks to attack the cooperative nature of oil production, marketing and distribution.

Trade promotion is to be designed as an integral part of government's poverty reduction programme. The policy holds that trade liberalization involves both import and export liberalization and intends to promote exports by:

- Enhancing the marketability of exportable through product diversification standardization and quality improvement;
- Strengthening and improving the institutional framework for providing better support services to exporters and export oriented industries;
- Establishing backward linkage between export oriented industries and primary sectors for the utilization of local raw materials;
- Attracting increased number of entrepreneurs for setting up export oriented industries and encourage them through the provision of suitable incentive packages, as well as appropriate human resource development programmes for the promotion of entrepreneurial and management skills in the context of a competitive international environment;
- Expanding and consolidating existing export markets as well as create new markets for Nigerian exports;
- Ensuring the removal of procedural and regulatory bottlenecks incompatible with the attainment of the objectives of a dynamic export promotion policy;

- ☆ Promoting programmes for developing export-oriented knowledge based resources, including computer software, electronic commerce, engineering and consultancy services;
- ☆ Diversifying and increasing export of high value added manufactured product that depend on the natural resources where Nigeria has comparative advantage;
- ☆ Encouraging the acquisition and adaptation of environment friendly technologies to ensure that Nigerian products meet the required international standards;
- ☆ Establishing border markets to promote legal trading across Nigeria's borders and so discourage smuggling.

The import policy is based on the progressive liberalisation of the trade regime thereby encouraging efficiency and competitiveness of domestic industries and eliminating pointless protection (Article 2.6). The policy endeavors to remove quantitative trade restriction measures in line with WTO in all traded goods and service while safeguarding the genuine interest of domestic producers against unfair trade practices. The policy will check the importation of injurious goods and services and thereby protect the social morality, plant and animal life and physical environment of Nigeria. The policy promises an efficient and transparent tax system, disciplined customs administration, effective commercial courts, improvements in infrastructure and the financial system, etc. Given that the economy is expected to be driven by private sector initiatives and competition, the policy foresees the development of a competition policy and an agency to deal with all forms of anti-competitive practices, mergers and acquisitions. It campaignes privatization, macroeconomic reforms especially in fiscal, monetary, taxation and tariff policies to effectively tackle the twin problems of high interest and inflation rates.

According to the policy, other strategies and measures to enhance the investment and general business environment include:

- ☆ Liberalisation that recognizes the need to protect domestic industries from unfair trade practices; promotion of conducive investment for foreign investment;
- ☆ Control of dumping and the establishment of the Anti Dumping Commission;
- ☆ Control of smuggling;
- ☆ Guarantee of fair and adequate returns on investments;
- ☆ Promote inter-sectoral linkage between trade and other sectors of the economy;
- ☆ Strengthening of the legal and regulatory frameworks of guidelines, standards, procedures and measures to manage bankruptcy, acquisition, mergers, intellectual property, integration of the informal sector into the economic mainstream, inspection and monitoring arrangements etc.

The policy gives special attention to agriculture, industry and service. For agriculture, the policy dealt with such critical issues as priority areas of export promotion, agriculture trade development especially in relation to the provision of enabling environment, industrial and commercial agriculture, liberal and efficient tariff system produce inspection and quality control and the establishment of agricultural trade information bureau. For industry on the other hand, the policy's main objective is to stimulate competition and promote investments in Nigeria with a bias towards the export market while also developing the skills and capacity to complete with imports in the domestic market. The policy also recognizes the inseparable link between trade, investments and industrial policy, but promises to pay priority attention to export promotion, new initiatives in trade policy, product tracking, creation of market niches, trade in technology and e-commerce. It also addresses research to industry linkage, anticipatory market development, transit trade, short-term capital financing for exporters and solid minerals development. Trade support infrastructure provisions of the trade policy to include the elimination of distortions and provision of international standard service to producers and exporters. The document envisages a drastic reduction in the domestic transaction costs to facilitate international competitiveness of export products from Nigeria. The provisions include trade information service, customs, quality certification and control, improvements in banking and insurance services, promotion of distributive trade, control of trade malpractices, weights and measures, provision of border free zones, trade documentation and procedures, bonded warehouse, among others. The policy further provides for export incentives including the export development fund, export expansion grant fund, duty drawback scheme, pioneer status, capital assets depreciation allowance, investment tax credit scheme. Other incentives include tax relief, interest income, rediscounting of short term bills, and retention of export proceeds, export credit guarantee and insurance scheme, manufacture-in-bond scheme, counter trade arrangements/buy back scheme. The policy also provides for free trade zones/export processing zones and trade statistic and data bank.

Section fourtheen deals with institutional and administrative framework for the implementation of the policy, it allocates duties among the various tiers of government and the private sector. The document itself acknowledges that this is a novel factor in trade policy design in Nigeria, as previous documents did not explicitly provide for such institutional framework. The section charges the Federal Government with the provision of general policy framework for the development of trade, coordination of data on trade and facilitation of research into all facets of trade; advise on tariff and pricing policy for agricultural produce, minerals, industrial products and services; promotion of the sale of exportable agricultural and industrial products; maintenance of a reasonable flow of resources into trade development by way of matching grants, subventions and investments projects and facilitating resource flow other sources; training of manpower (inspectors, export developers and import management experts) for trade development; and promotion of appropriate technology in preservation, processing, packing, quality control, standardization, storage, financing, risk-bearing, market intelligence, transportation and market structures, through the development of prototypes and pilot projects as a follow-up to research inventions. State Governments are charged with the promotion of primary production of all items of export-oriented agricultural and industrial products; training of manpower for trade growth and development; establishment of appropriate institutions for purposes of price stabilization; and investments in access road, water supplies and electricity in the localities where these tradable products for purposes of price stabilization; and investments in access road, water supplies and electricity in the localities where these tradable products are obtained. Local Governments Councils are expected to support states and progressively take over such functions for the States as the provision of an effective trade extension service; provision of site infrastructure; mobilization of traders for accelerated trade development through cooperative organizations and community associations; provision of assistance in data collection at the primary level; and construction, management and maintenance of local markets to enhance trade.

The private sector on the other hand is the major driver of trade and so is expected to drive the realization of the objectives of the trade policy. Its leading roles are in the areas investment in all aspects of production of tradable commodities; storage, preservation, packaging, processing and other marketing functions; input supply and credit disbursement; provision of certain types of tax-deductible site infrastructure; and support for research in all aspects of trade development; and organization of training workshops, trade fairs and trade missions to promote innovation and wider participation of their members.

National Focal Point on Multilateral Trading Matters;

- ☆ Trade Policy Advisory Committee to coordinate policy formulation and negotiations among relevant government ministries;
- ☆ Executive Secretariat;
- ☆ Trade and Competition Commission;
- ☆ National Council on Commerce;
- ☆ Bankruptcy Commission.

The section also makes provisions for other support bodies both in the public and private sector. For example, the policy also envisages commercial officer services; trade related bodies like the Nigerian Export Promotion Council, Nigerian Export Processing Zones Authority, Nigerian Investment Promotion Council, Standards Organization of Nigeria etc. Also stakeholder consultations are expected to make inputs into trade policy while an Economic Policy Coordinating Committee is to facilitate the realization of government's economic and fiscal goals as it relates to trade policy. The Manufacturers Association of Nigeria; National Association of Chambers of Commerce, Industries, Mines andAgriculture and other private sector operators are expected to\play key roles in trade policy implementation. The Trade Policy ends with an action plan and implementation strategy, which identifies problems, recommends solutions/implementation strategies and incorporates a time frame as well as the implementing agencies.

15

Gap in the Existing Trade Policy

15.1. Gap in the Existing Trade Policy

Some of the straightforwardly identifiable gaps in the existing trade policy include the first objective of the policy is to integrate the Nigerian economy into the global economy through the establishment of a liberal market economy. This idea could be deceptive. What is essential is not a link of the Nigeria economy to the world economy as such, (because by the contemporary trading relations, Nigeria is already linked) however, a framework for the reform of the terms, processes and procedures that notify the linkage. This will guarantee not just free trade, but fair trade as well as sustainable development for the country. The objective of fairness in trade habitually link up to the poverty question with the probability that fair trade will reduce poverty.

There is modest if any effectual link between trade and industrial policies in Nigeria, Such link should be able to deal with the effects of the latter on the performance of the former and how policymakers propose to deal with the problems of poor hold up infrastructure that elevate the cost of doing business. This is connected to the question of diversification of the nation's productive base away from oil. Incentive and packages to attract investments in new refineries and petrochemical plants would have been an absolute way of adding value to oil exports and providing linkages of the oil sector to other sectors of the economy. The fact that five years after the policy, Nigeria is still importing refined petroleum products is an indication in the direction of such harmonization failure. The projected diversification may never happen at this rate. The economy's reliance on oil has risen over the last five years of the implementation of the policy.

These discrepancies show up in operation record where the policy has again scored very low. This is evident in the following areas;20

- While under trade endorsement, the policy projected to reduce poverty by 50% calculated from the policy point of reference year in 2005 and raise employment between 60 per cent and 70 per cent. It is important to note here that both poverty and unemployment have worsened.
- Some important laws, policies and agencies like Anti-trust and competition laws, anti-dumping agency and bankruptcy commission contempt under the policy for trade facilitation have not been established.
- Already established agencies like the Standards Organization of Nigeria are yet to meet their mandates of enforcing standards. The Export Processing Zones Authority and Nigeria Investment Promotion Council are also yet to attract sufficient investments in their respective area of authority.
- The projected bottom-up empowerment approach of stakeholder and civil society to demand accountability are yet to be realized as policy implementation has barely sought their inputs and there is no synchronized support from government to help them build critical analytical skills to assess trade policy implementation.
- The macroeconomic reforms necessary for the successful implementation of trade policy have not been placed.
- There has also been a discrepancy in the implementations of the provisions of the trade policy itself. For example, between 2000 and 2004, the country has moved from pre-shipment inspection to destination inspection and is in the process of reverting to pre-shipment inspection.

Governments at all time in Nigeria have been very involved in the use of instruments of trade policy. The instruments in use and degree of their employment have also varied extensively depending on the objective purpose sought to be exploited by the government. Some of the most common objectives over time comprises of balance of payment stability, utmost use of government revenue, encouragement of economic growth, improved employment of human and material resources, in addition to fortification of infant industries, which although should be an intermediate aim, frequently ends up being an end itself. Simultaneously, some of the most regularly used instruments consist of import prohibition/licensing, export incentives (rebates, tax holidays, export processing zones etc), exchange rate and tariff. Of these, modifications in tariffs and connected taxes and waiver have been inclined to be the most important. Nevertheless, historically; the success of the use of trade policy instruments by the Nigerian government has been under severe scrutiny. There is deep belief that the policy instruments have not been generally effective. This owes to a number of reasons. First, Nigeria's trade and tariff policy have been very capricious and the rate of instrument change has been extremely high leading to a general loss of confidence by major operators and stakeholders in the instruments. It is not unusual for investors to pay no attention to the chances created by a significance prohibition on the indulgent that the prohibition will not linger long enough to license advantageous investment in the sector. Moreover, the efficiency of the implementation of a given policy is very low. Real tariffs have constantly differed from published tariffs while efficiency of collection of even the real tariff had at times dropped lower than 50 per cent. The technique of administration of waivers has always left much to be considered necessary leaving collected revenue often differing much from collectible revenue.

There is also the conviction among many stakeholders that import prohibitions only deepen customs and other public official through corruption as the prohibited items will inevitably enter the country anyway. This is in addition the poor merging of either the views of key stakeholders or even the result of meticulous research in the adoption of trade policy instruments such that often a times there has been very little if any stakeholder buying into the programmes. Exchange rate policies have been carried out more by rule of thumb and public opinion than with empirical evidence. A broad range of export motivation proposals has failed to make up for anti-export prejudice of the average Nigerian producer while complementary export and production plans and documentation remain as unwieldy as possible. Support infrastructure is literally absent and the 1990s saw minute investment in infrastructure that should support trade policy. But significantly too, the country has not been stable in the preservation of a clear cut trade policy path. While the overall trend has been towards larger liberalization, there have been a number of policy turnarounds and *ad-hocism* connected with trade policy and which tended to give stakeholders the notion that the government is unserious. The current trade policy refers to various trade facilitation and support measure to advance on the quality and success of trade policy instruments. This include stable exchange rate system, successful competition and anti-trust policy, improved harmonized export incentives and institutions for overseeing them, harmonizing infrastructure provision among others. Nonetheless, a lot of of these have not come to past. But crucially, there is need for strategic inclusion of stakeholders in the design and implementation of trade policy. The implementation of any trade policy instrument also needs to be guided by empirical evidence and not just the emotions of a few individuals with access to power.

There is in fact not much amiss with the present export promotion policies and competition as written in the trade policy document. But there is need for action to improve these activities through suitable incentives. Some of the steps already taken by the present administration particularly regarding the introduction of larger competition through privatization and reducing documentation troubles connected with export rebates and investment, these need to be permanent and institutionalized.

15.2. The Trade Policy and the National Economic Empowerment and Development Strategy (NEEDS): An Overview

In 2004, the Federal Government of Nigeria changed its reform programme to the National Economic Empowerment and Development Strategy otherwise known as NEEDS. NEEDS came as a response to the need to re-evaluate the policy process, re-evaluate the operation environment, remove obstructions to growth, and speedup economic development in Nigeria. NEEDS 'is Nigeria's home-grown poverty reduction strategy (PRSP) founded on clear vision, sound values, and enduring principles. It is a medium term plan, but which derives from the country's long-term goals of poverty reduction, wealth creation, employment generation and value re-orientation.' It is predominantly a reasoned response to the challenges of underdevelopment and endemic poverty that has historically been the lot of Nigeria. It rests on four key strategies: Reforming government and its institutions; Growing the private sector; Implementing a social charter; and Value change and reorientation of the people towards an enduring African value system. Generally NEEDS commits to improve efficiency, and productivity, reduce production costs, nurture entrepreneurship and enhance a competitive marketplace. In more specific term, NEEDS tries to achieve wealth creation, employment generation, poverty reduction and value reorientation.

Within the scope of its public sector reform programme, government intends to restructure, right-size, re-professionalize and strengthen government and public institutions to bring forth effective service to the people. There is also the goal of eradicating waste and inefficiency, and create resourceful space for investment in infrastructure and social service by Government, by fighting corruption and promotion of the rule of law. NEEDS also promises to develop the private sector by enhancing its competence to invest and manage businesses. The key elements of this strategy include privatization, de-regulation and liberalization programme targeted to shrivel the sphere of the public sector and hold up the private sector; infrastructure development especially electricity (power) and transportation; unequivocal sectoral plan for agriculture, industry/SMSEs; service (especially tourism, art Under NEEDS, the social charter is defined in terms of the contract between government and people of Nigeria in a formal agent-principal relationship and where the ultimate objectives of governance is the improvement of the welfare of residents and culture, and information/communication technology), oil and gas, and solid minerals. It also identifies the need for long-term capital mobilization; institution of appropriate regulatory framework; a logical and dependable trade policy and regional/global integration regime; and specific interventions to encourage the development of some sector, promotion of industrial parks as well as research and development

The social charter deals with specific programmes for development of health, education, employment, poverty-reduction, security and participation. Education is treated as a major instrument of empowerment while reforms in the health care delivery system will be oriented sector towards, effective health care delivery system with a view to combating preventable diseases like malaria and stemming the spread of HIV/AIDS. There are also programmes directed at youth re-orientation and employment. Pension scheme is also up for review while housing and mortgage development is to be led by the private sector but with substantial government input. These provisions (including investments in water, both as a social facility and for irrigation) are to be guided by the need for environmental sustainability. The social charter also seeks inclusiveness and empowerment with programmes designed to give voice to the weak and vulnerable culminating in an 'affirmative action for a minimum of 30 per cent representation for women in all aspect of national life wherever feasible'. Value re-orientation seeks to address issues of corruption and rent-seeking and to give relevance and prominence to industry as a means of livelihood. Each of the earlier three components of the reform programme is expected to make inputs into the value re-orientation programme. A persistent decimal in the policy making scenery in Nigeria has been policy implementation gap. Consecutive processes and analyses have referred to this as the curse of policies in Nigeria. It is often held in many quarters that Nigeria does not lack good ideas, but often runs short in implementation of designed programmes. NEEDS foresees and tries to address this by providing a coordination, reporting and implementation framework that is based on a realistic appraisal and strengthening of existing institutions. NEEDS is also refreshingly innovative in the design of a framework for collaboration and coordination with all tiers and agencies of government having substantial contributions t make with effective complementation from the private sector.

NEEDS appears to depart from previous development strategies in its inclusion of the human development agenda and empowerment. Specific human development issues form the critical mass for the success of the reforms process, because, it is premised on the assumption that no economic policy is successful until it has translated into tangible benefits for the people. NEEDS incorporates the principles of balances development in the planning and management of the nation's human and physical development so as to ensure it reflects the aspirations of the diverse interests in the nation. While it did not disaggregate development initiatives along the cross cutting issue of ensuring gender equity, as a strategy to redress structural imbalance that have continued to perpetuate poverty, discrimination and limited access to opportunities, it provided for an affirmative action of 30 per cent representation for women. NEEDS' overwhelming objective is the reining in of government. The NEEDS developmental structure might involve some temporary shocks especially to the vulnerable. But how 'temporary' such shocks will be depends critically on the commitment of government to the provisions of the social charter and the targets specifies there. For example, NEEDS provides for 'right-sizing' government which might entail shuffles in public sector employment. Meanwhile, there are no guarantees that anticipated private sector engagement will automatically reabsorb such dislocated workers or provide buffers against the pains of adjustment. In general, the NEEDS macroeconomic framework and approach is conventional, framed on three fundamental strategies; strengthening and re-orienting government and governance, growing the private sector and implementing a social charter. However, there could be tensions among these three objective/strategies, at least in the short run. For NEEDS promises 7 million new jobs by the year 2007, increased immunization coverage to 60%, increased access to potable water by 70% of the population, increased School enrolment rates especially for girls and increased adult literacy rate to at least 65%. It als promises to significantly improve sanitation. example, the booming and well articulated social charter could in the short run be hindered by an attempt to 'force' government to act, as when restraining in the public sector could mean downsizing.

15.3. The Nigeria Trade Policy and Needs: An Appraisal of Convergence

The Nigerian trade policy identifies that its effectiveness is inextricably linked to sound domestic policies, which are implemented in a flexible manner. Economist in Nigeria argue that it is encouraging to note a major degree of convergence between the 'spirit' of the trade policy and NEEDS especially in the following areas:

a. **Complementary Roles of Government and the Private Sector:** Whether in the NEEDS or the Trade Policy documents, efforts were made to clearly redefine the role of government as that of a facilitator that provides the facilitating environment necessary for the growth of a competitive private sector. The private sector, on the other hand, is expected to take the lead in the productive sector of the economy by taking advantage of abounding opportunities in the domestic, regional and global markets. Both government and the private sector endorse the accomplishment of accelerated growth and development through trade within the public-private partnership concept. Both documents advocate targeted interventions by government to strengthen weak and susceptible sectors and stress the predictability of effective partnership between both sectors of the economy in the up-and-coming economic reconstruction.

b. The Role of International Economic Developments in Domestic Policy Design:

   Both the NEEDS and the Trade Policy recognize the significant impact that global economic developments have on national policies. Certainly, one is tempted to describe both the NEEDS and the Trade Policy as responses to changes in the global economic community. The integration of Nigeria into the global economy is at the heart of NEEDS. For both documents, the starting point is the initial integration with the rest of the countries of the ECOWAS sub region.

c. **Diversification of the Productive and Export Bases:** Both the trade policy and NEEDS recognize the need for the diversification of the productive and export bases of the economy away from oil and to promote market-oriented private sector driven economic development. Both documents concur on the attainment of growth through the promotion and diversification of exports in oil and non-oil markets. In this regard, there is an agreement on the need to take on nationally harmonized strategies to develop significant sectors like agriculture, small and medium enterprises, manufacturing, solid mineral and services.

d. **Investment in Infrastructure and Reduction in Cost of Doing Business:** Another point of agreement Agu and Oduh poses is on the need for vast investments in infrastructure, especially electricity, transport and water resources. Within the West African sub-region, Nigeria ranks as having one of the poorest infrastructure and which capacity has been outshone by population growth, especially in the urban centres. This further aggravates the present costs of business. There is agreement on the need to 'lower the cost of doing business, in addition to reducing policy-related cost and risk corruption, red tape and administrative barriers to business including a weak legal system; inadequate protect of property rights; inadequate enforcement of tariffs; dumping of fake and sub-standard products; poor payment systems, and the uncertainties associated with exchange rate and prices.' NEEDS also predicates trade and investment on the rule of law and security through more implicitly state in the Trade Policy Document.

e. **Sector-Specific Issues:** In terms of sectoral emphasis and the description of growth drivers, there is much agreement between NEEDS and present Trade Policy.

   Both underline and articulate interest in the endorsement of agriculture for food security and particularly in respect of the proportion of the population that are employed there. The trade policy recognizes Nigeria as an agro-based economy and so targets the production of basic food products, in order to ensure food security and make Nigeria a net exporter of food. NEEDS also emphasize the place of agriculture in the following way: "Agriculture remains the mainstay of Nigeria's economy. In addition to contributing the largest share of GDP, it is the largest non- oil export earner, the largest employer of labour, and a key contributor to wealth about 60 per cent of the Nigerian population are farmers, engaging in subsistence agriculture and over 65% of these farmers are women and children. Creation and poverty alleviation, as a large percentage of the population derives its income from agriculture and related activities"

Nonetheless, Agu and Oduh notice that there are some discrepancies both in the position and spirit of the two policy documents particularly in respect of the choice of selective import restriction versus openness. Both documents are incredibly outward oriented and receptive to developments in the economy. To this end, while the Trade Policy was forceful outright liberalization of both the export and import sectors, NEEDS emphasized the need for selective import restriction and aggressive export promotion as a strategy to ensure orderly re-structuring of the industrial programme. This deviation is by no means small bearing in mind the structure of economic activities in Nigeria. Being a policy document, the NEEDS could not go down to details in describing what constitutes 'selective import restriction' and the modalities of its implementation. In a situation where the trade policy is emphatic about liberalization and the full implementation of all international protocol to which Nigeria is signatory, there is the propensity for divergences in implementation of programmes proposed by the two policy documents.

15.4. Nigeria's Trade Policy and the Livelihood of the Poor: Emerging Realities

So far, this paper has attempted to reveal a correlation between increasing occurrence of poverty and segregation on the one hand and increased liberalization of trade on the other. This section presents a summary of the major issues raised thus far. Trade for every country has two major aspects - domestic and international. Conceptually and operationally, the demands of both trades are not exactly the same particularly in a developing country context. There could thus be some friction, the extent of which depends on the level of development of the country involved, the size of the supportive infrastructure and environment for promoting both trades and the political/other interest pressures supporting the promotion of either trade. Managing the linkage between the two sets of trade could be difficult and so far, the existing Nigeria trade policy did not demonstrate effectiveness in managing it. The 2002 trade policy is excessively outward-oriented. The document demonstrates undue eagerness to satisfy the demands of international trade. The policy pays unwarranted attention to issue of global importance but does not show equivalent commitment to the needs of home trade which holds great potentials for employment and poverty reduction. The policy fails to demonstrate understanding of the necessary linkages between the international trade demands and domestic economic growth. There is the supposition that liberalization by itself leads to growth and poverty reduction, but as has been shown in the body of this paper, this is not necessarily so. The significant transmission mechanism or linkages between outward oriented trade policy and domestic growth was not scrupulously obtained or shown in the document.

Export promotion and import liberalization are all important and realistic policy objectives. But the character and structure of such promotion and indeed driving force behind these policies matter. It is essential that the synergy between such outward promotion programmes and national output growth and development of applicable infrastructure be demonstrated. Agu and Oduh in their report expressed their fear regarding domestic output and trade may continue to suffer if this is not attended to. The current goods and services distribution network in the country is anything but desirable. Available resources could either be used for large export promotion activities or channelled through policy into the hands of artisan and operators in the informal sector where poverty is predominant. Import liberalization on the other hand pressurizes domestic producer particularly in view of the irregular infrastructure and high cost of doing business in the country, reduces their competitiveness and leads to further unemployment which worsens poverty. These are demonstrable points proven by over two decades of experience in trade liberalization.

15.4.1. What then is the Way Forward?

Given the level of development of Nigeria, it is recommended that trade policy is merged with industrial policy. It is argued that Nigeria does not need two separate documents to deal with trade and industrial development. Such document will have to pay attention to issues affecting domestic production and distribution.

'Nigeria's federal character can be very advantageous in the respect as the National Economic Council becomes the melting pot for coordination of state efforts in complementary real sector development while the federal government provides overall macroeconomic framework and matching grants to make this possible. Agu and Oduh in their report further argue that there is need for 'strategic liberalization' and that quick and unrestrained liberalization of the Imports sub sector is unnecessary and could be damaging. True enough, such strategic liberalization programme needs to be steady and selective in sectors to liberalize. For example, while the country needs capital and raw materials to promote domestic output, there is little need for liberalization of the imports of food, light manufactures and/or textiles. Likewise, export promotion is desirable, but it will be important for policy to be specific on which sectors priority attention shall be given and why. On the other, it is feared that unrestrained openness and liberalization of imports, would affect the income of strategic sectors and possibly wipe away the little competitive advantage that they currently enjoy. Under such circumstances, the productive base that could support exports will be further eroded.

The lack of effective structure for trade policy implementation in Nigeria is one of the major factors for the ad hoc nature of import bans, indiscretion in the movement of tariffs and exemptions as well as the diversity and challenges in the various trade protocols to which Nigeria is signatory or still intends to be signatory. These set of problems can only be addressed by proper institutionalization of the processes of tariff collection, reporting lines and granting of exemptions, trade negotiation as well as economic cooperation between Nigeria and other countries. Agriculture and manufacturing continue to decrease and minimal efforts are made to develop competitive advantage in these sectors. Given their huge employment for the majority of the poor, this stance has implications for the overall effectiveness of trade policy as an instrument of poverty reduction in Nigeria. Specifically, while the impact of trade policy has been most felt by the poor, understanding of such impact and implications is somewhat vague. This owes to the woolly treatment that trade policy gives to agricultural and industrial development strategies. There is very little attempt at explicitly treating trade in agricultural inputs and output, linkage of agricultural products with industrial users, exports of agricultural products to regional markets, stabilization of agricultural product prices, and subsidization of specific inputs like land-clearing, if necessary. More so, the link between agricultural and trade policies is weak. There is little (if any) complementarily between trade and agricultural policies and the strategies proposed by one are often at variance with that proposed by the other. Finally, both the agricultural and industrial sectors need large amounts of infrastructure development. This is both a policy design and implementation challenge. Liberalization on its own means nothing. Certainly, for a country like Nigeria with most basic support infrastructure not working, liberalization may imply negative growth. It has been acknowledged that a number of complementary policies and infrastructure is needed if trade policies are to make the required impact on real sector growth. As earlier noted, most of such complementary conditions for effective realization of the goals of the policy (low interest and inflation rates, reduced debt burden, etc.) are yet to be realized. Neither has infrastructure worked. However, trade liberalization has ensued at striking rates. Thus, there is need to bind in the strategic liberalization programme with substantial and unrelenting investment in infrastructure.

15.5. Nigerial Balance of Trade

Nigeria reported a trade surplus equivalent to 9.2 Billion USD in December of 2011. Historically, from 2002 until 2011, Nigeria Balance of Trade averaged 2.5800 Billion USD reaching an all time high of 14.5900 Billion USD in October of 2011 and a record low of -3.8200 Billion USD in March of 2011. An export of commodities (oil and natural gas) is the main factor behind Nigeria's growth and accounts for more than 95% of total exports. Nigeria's main exports partners are: USA (30% of total in 2009), Equatorial Guinea (8%), Brazil (6.6%), France (6%) and India (6%). Nigeria imports mainly: industrial supplies (32% of total), transport equipment and parts (23%), capital goods (24%), food and beverage (11%) and consumer goods. Main import partners are: China (17% of total), Albania (11.3%), United States (7.5%), France and Belgium. This page includes a chart with historical data for Nigeria Balance of Trade.

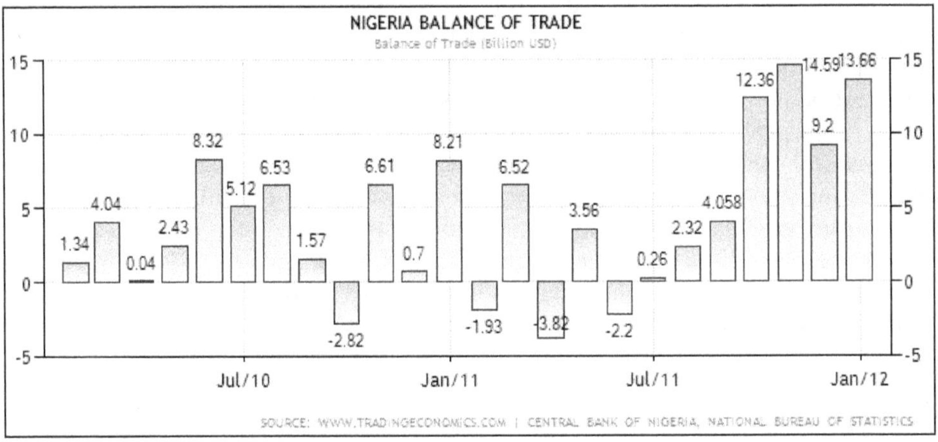

Indigenous insurance companies have alleged that the country loses about N45 billion annually due to the preference given to foreign ship owners and their choice of insurers over the indigenous companies in the lifting and importation of fuel. They further alleged that the Nigerian National Petroleum Corporation, NNPC, deliberately sidelined insurance companies from participating in the insurance of imported fuel both locally and internationally. According to them, the NNPC sets unnecessary bulwarks that made it impossible for Nigerian insurance companies to participate in the insurance of imported oil business, thus shortchanging Nigeria as the country loses as much as N3.7 billion monthly in freight insurance that it should be earning. This is happening despite the existence of the Cabotage law in Nigeria. Last year, Mr. Fola Daniel, Commissioner for Insurance, said that the National Insurance Commission (NAICOM) would not permit risks to be underwritten offshore unless local capacity has been exhausted. Daniel also said the Nigerian Oil and Gas Content Development Act 2010 desired that local insurers and re-insurers will be able to write a minimum of 45 per cent of oil risks in 2007, and 70 per cent at the end of last year, but lamented that only 33 per cent of the risks in the sector was retained locally.

He attributed the development to what he called absence of an enabling law in the past to back up the policy initiatives, adding that with the coming into effect of the Local Content Act, in accordance with Insurance Act 2003 and the National Insurance Act 1997, and the release of the guidelines for oil and gas business in Nigeria, 100 per cent, 70 per cent and 40 per cent of all life, non-life and marine insurance risks in the Nigerian oil and gas industry must be placed with insurers in Nigeria. According to Daniel: "No risk can be placed off-shore without the written approval of the Commission, which is required to ensure that Nigerian local capacity has been fully exhausted."

When Financial Vanguard contacted the NNPC spokesperson, Mr. Levi Ajuonumah, for comments, he denied the allegation, saying that all the vessels coming into the country were insured but refused to divulge if they were locally or internationally insured. An insurance expert who prefers to be anonymous said that all these notwithstanding; the insurance industry has been shortchanged all along. According to him, the industry was shortchanged when the exact quantity of petrol on which subsidy was being charged were not known as there were no verifiable records of importation into the country. The industry was also shortchanged when the Nigerian National Petroleum Corporation, NNPC, refused to divulge its volume of imports and was said not to have paid any duty to the Federal Government on imported petrol since 2002 owing to a presidential waiver. Prior to 2002, the corporation allegedly owed the Nigeria Customs Service over N46 billion as accrued duty on petrol.

The insurance industry was shortchanged yet again when mother vessels berthed off-shore to discharge products instead of going to designated ports and while on the high seas, the vessels discharged products into smaller vessels for onward delivery to the ports and because Customs personnel were barred from examining offshore vessels, they could not determine the exact quantity of products the vessels brought into the country. It will be recalled that the Customs and Excise Management Act provides that vessels from foreign countries must berth at the ports of destination, not at unapproved locations.

According to Paschal Egerue, Managing Director of Mainstreet Insurance Brokers, insurance fraud include non-declaration of transactions to NAICOM, fronting for overseas insurance companies, bearing uninsured risks or inappropriately reinsured risks, premium loading or premium under-declaration, round tripping, among others. Egerue stressed that all fuel importation ought to have been duly insured by the oil majors. Chairman of Nigerian Insurers Association, NIA, Mr. Olusola Ladipo-Ajayi, said that insurance operators do not have anything to show that goods worth the amount claimed by officials in the oil sector have been insured in the Nigerian market as even government officials could not agree on figures.

Ladipo-Ajayi said: "The Central Bank of Nigeria gives a figure different from what the NNPC gave and different from what the Ministry of Finance gives.

Everything is just working at cross purposes but we do not have any figure to come near anything that anyone of them has said." Meanwhile, Egerue noted that the Local Content Act was enacted after the debate as to the capacity of the nation's insurance industry to absorb the risks of the Nigerian oil and gas upstream and downstream sectors and NAICOM has since moved in to enforce the sections of the Insurance Act relating to oil and gas insurance and reinsurance in Nigeria. According to him, there is a school of thought that principally sees insurance transaction in oil and gas sector as ridden with fraud. The Nigerian insurance industry has been losing billions of dollars to these occurrences.

In his words: "The fuel subsidy argument has therefore brought to the fore the likely quantum of these frauds. The crucial question is, since this subsidy may also encompass insurance premium reimbursement, who computes it and who has been taking it? In any case, how has the premium for the massive marine cargo insurance on these fuel importations been handled all these years?"Most of the ship-loads of fuel are insured on clause C at very ridiculous rates while possible reimbursements may be calculated on clause A on exaggerated rate basis. There are also the experiences of incessant certificate cancellations and substitutions as these importers play on the frequency of importation to cheat the insurance industry. These irregularities are, however, cleverly reconstructed whenever a loss occurs. Thus, insurance companies are seen to exist only to bear losses and not to earn income from this huge racket. "Mr. Yemi Soladoye, an insurance consultant said that the law actually stipulates that all imports in Nigeria must be insured through a local underwriter, adding: " If you break that down, it is not saying all imports must be insured but if an import will be insured, it must be through a local underwriter."

Managing Director of FBN Life Assurance Company Limited, Mr. Val Ojumah, stated that the law says all import and export into Nigeria must be insured by Nigerian insurance industry, but since the government refuses to do so, insurers cannot force it to comply. He said: "If they decide to disobey the law, it is not in our place to actually enforce the law. We have the police and we have the government to enforce the law." According to him, most government imports are not declared to insurers because the government that brings in such imports does not bother to insure them."Now if you take a vessel laden with PMS from Port-Harcourt, turn around on the high sea and bring it to Lagos and record it as import, do you want me to insure that? That could not be declared to me. Given the fact that in some of these cases they didn't buy insurance and personally, I believe they wouldn't buy insurance because insurance operators cannot be part of what they were doing. Insurance cannot insure a fraudulent situation." Ladipo-Ajayi stressed that the insurance sector did not benefit from the huge figures being bandied about. He said: "There is a lot of information that came out of the fuel subsidy probe. However, with that amount of fuel importation, if the Nigerian law that makes it compulsory for the importation of goods into Nigeria to be insured has been observed, you would have seen it in our reports. But I don't think we have anything to show that goods worth that amount have been insured in the Nigerian market. Even the government officials in charge cannot agree on figures."

16

Foreign Trade Barriers in Nigeria

16.1. Foreign Trade Barriers in Nigeria

(a) Nigeria Trade Summary

The U.S. goods trade deficit with Nigeria was $34 billion in 2008, an increase of $4 billion from $30 billion in 2007. U.S. goods exports in 2008 were $4.1 billion, up 47.7 per cent from the previous year.

Corresponding U.S. imports from Nigeria were $38.1 billion, up 16.2 per cent. Nigeria is currently the 44th largest export market for U.S. goods.

The stock of U.S. foreign direct investment (FDI) in Nigeria was $190 million in 2006 (latest data available).

(b) Import Policies Tarriffs

In September 2008, the Nigerian government issued the 2008-2012 Common External Tariff (CET) Book that harmonizes its tariffs with its West African neighbors' under the Economic Community of West African States (ECOWAS) Common External Tariff (CET). Nigeria has been partially implementing the CET since 2005. The new tariff regime has five tariff bands and import duties have been reduced on a number of items, such as rice, cigars, and manufactured tobacco. The five CET tariff bands are: zero duty on capital goods, machinery, and essential drugs not produced locally; 5 per cent on imported raw materials; 10 per cent on intermediate goods; 20 per cent on finished goods; and 35 per cent on goods in certain sectors. The fifth band is still under negotiation among member countries. Adoption of the CET is part of ongoing economic reforms aimed at improving Nigeria's trade and investment environment and harmonization of economic policies in the sub-region. There remains resistance within the Nigerian government and the private sector to deepen trade reforms.

Companies state that high tariffs, nontransparent valuation procedures, frequent policy changes and unclear interpretations by the Nigerian Customs Service (NCS) continue to make importing difficult and expensive, and often create bottlenecks for commercial activities. Some importers complain that tariffs are excessively high and that the Nigerian government sometimes uses arbitrary reference prices for valuation purposes. This problem is aggravated by Nigeria's dependence on imported raw materials and finished goods and affects both foreign and domestic manufacturers. Reportedly, many importers resort to undervaluing and smuggling to avoid paying full tariffs. Transparent and proper implementation of the CET would be an important step toward resolving most of these problems.

(c) Non Tariff Measures

Though the government continues to ban certain imports, citing the need to protect local industries, the new tariff book reduces the number of items on the import prohibition list from 44 to 26. Items removed from the list include corn, sorghum, millet, wheat flour, crude vegetable oil, biscuits, sugar confectioneries (including white chocolate), fresh and dried fruit, flowers (both fresh and plastic), toothpaste, envelopes, diaries, greeting cards, exercise books, bentonites, barites, calendars, cutlasses, axes, pick axes, spades, shovels, fully built mudguards, wheel barrows, and electric generating sound proof casings.

Items remaining on the import prohibition list include: bird's eggs, cocoa butter, powder and cakes, pork, beef, live birds, frozen poultry, refined vegetable oil and fats, cassava, bottled water, spaghetti, noodles,Fruit juice in retail packs, nonalcoholic beverages (excluding energy drinks), certain textile products, and bagged cement. Companies were awarded concessions to import bagged cement for a limited time to bridge supply gaps. The cement concessions terminated at the end of December 2008, but the companies have applied for an extension.

(d) Customs Administration

Nigeria practices a destination inspection policy for imports. Under this policy, all imports are inspected upon arrival into Nigeria, rather than at the ports of origin. Nigeria port practices continue to present major obstacles to trade. The country's long list of items prohibited for import, coupled with incorrect declaration of goods by importers, result in 95 per cent of containers being physically examined. This delays the clearing process and increases costs. Nigeria's uneven application of import and labeling regulations make importing high-value perishable products difficult. Disputes between Nigerian agencies over the interpretation of regulations often cause delays, and frequent changes in customs guidelines slow the movement of goods through Nigerian ports. Importers report erratic application of customs regulations, lengthy clearance procedures, high berthing and unloading costs, and corruption. These factors can contribute to product deterioration and may result in significant losses for importers of perishable goods.

Realizing that delays at the ports significantly increase the cost of doing business in Nigeria, the Nigerian government plans to implement a 48 hour cargo clearance policy at the ports. Roads coming in and out of the ports are decaying, and overuse results in around-the-clock traffic congestion. There is no rail system transporting freight in and out of ports. As such, despite the 48 hour custom clearance policy, congestion leads to ships queuing up to berth at cargo terminals and containers waiting to be transported out of the ports. The bottlenecks resulting from the lack of infrastructure at and around the ports affect the level of efficiency at which goods can be processed for import. Currently over 15 agencies are represented at the ports. In a bid to achieve the 48 hour cargo clearance target at the ports, the Nigerian government plans to withdraw all agencies, except customs, from the ports and improve the technical capacity of customs to handle special cargos through continuous training of personnel. There are also plans to automate all customs payments.

Export Subsidies and other Export Promotion Programs

Nigeria's government administers various export incentive programs such as tax concessions, export development funds, capital asset depreciation allowances, and foreign currency retention programs in addition to operating Free Trade Zones and Export Processing Zones. According to the 2008-2012 CET Book, most concessions, waivers or exemptions have been stopped; however, the Nigerian Export Promotion Council will continue to implement the Export Expansion Grant scheme to improve non-oil export performance.

16.2. Standards, Testing, Labelling and Certification

Rules concerning sanitary and phytosanitary standards, testing, and labeling are well defined, but bureaucratic hurdles slow the import approval process. Regardless of origin, Nigeria requires that all food, drug, cosmetic, and pesticide imports be accompanied by certificates of analysis from manufacturers and appropriate national authorities; and specified animal products, plants, seeds, and soils must be accompanied by proper inspection certificates. By law, items entering Nigeria must be labeled exclusively in the metric system. U.S. producers and exporters note that relabeling goods to meet this requirement is expensive and limits U.S. exports to Nigeria. The NCS is charged with preventing the entry of products with dual or multiple markings, but such items are often found in Nigerian markets. The National Agency for Food and Drug Administration and Control (NAFDAC) is charged with protecting Nigerian consumers from fraudulent or unhealthy products. The agency continues to focus special attention on eliminating the illicit importation of counterfeit and expired pharmaceuticals, particularly from East and South Asia. NAFDAC's limited capacity for carrying out inspections and testing contributes to what critics have characterized as an occasionally heavy-handed or arbitrary approach to regulatory enforcement which sometimes leads to delays in clearance of legitimate food imports.

Although Nigeria has no laws governing agricultural biotechnology or bio safety, the government is generally supportive of biotechnology. The Federal Ministry of Environment has presented draft Bio safety legislation to the National

Council on Environment, the highest decision-making body on environmental issues. If approved, the legislation will be sent to the National Executive Council of Ministers for ratification and then to the National Assembly for its consideration. The draft bill generally portrays products of biotechnology as safe for animal and human consumption; however, it includes a mandatory labeling requirement.

16.2.1. Government Procurement

The government has made modest progress on its pledge to conduct an open and competitive bidding process for government procurement. The Public Procurement Act, which was signed into law in June2007, established the Bureau of Public Procurement (BPP) in place of the Budget Monitoring and Price Intelligence Unit. The public procurement reforms are aimed at ensuring that the procurement process for public projects adheres to international standards for competitive bidding. The BPP acts as a clearinghouse for government contracts and monitors the implementation of projects to ensure compliance with contract terms and budgetary restrictions. Procurement above 50 million naira ($419,000) is subject to review by the BPP. The 36 state governments have also agreed to enact the Public Procurement Act in their respective states. Foreign companies incorporated in Nigeria receive national treatment in government procurement; government tenders are published in local newspapers; and a "tenders" journal is sold at local newspaper outlets. U.S. companies have won government contracts in several sectors. Unfortunately, many companies that have won contracts have subsequently had difficulty getting paid, often as a result of delays in the national budget process.

The National Petroleum Investment and Management Services agency's approval is required for all procurement in the energy sector with a value above $500,000. Approval processes are slow and can significantly increase the time and resources required for a given project. Nigeria is not a signatory to the WTO Agreement on General Procurement.

16.2.2. Intellectual Property Rights (IPR) Protection

Nigeria is a party to the World Intellectual Property Organization (WIPO) Convention, the Berne Convention for the Protection of Literary and Artistic Works, the Paris convention for the Protection of Industrial Property, the Patent cooperation Treaty, and the Patent Law Treaty. Nigeria has also signed the WIPO Copyright Treaty and the WIPO Performances and Phonograms Treaty. Legislation intended to establish a legal framework for an IPR system consistent with WTO obligations has been pending in the National Assembly for several years.

The government's lack of institutional capacity to address IPR issues is major constraints to enforcement. Relevant Nigerian institutions suffer from low morale, poor training, and limited resources. DespiteNigeria's active participation in the conventions cited above and growing interest among Nigerians in seeing their intellectual property protected, piracy remains a problem. Counterfeit automotive parts, pharmaceuticals, business and entertainment software, music and video recordings, and other consumer goods are sold openly, and piracy of books and optical disc products is also a problem. Industry reports contend that intellectual property infringers from other countries appear increasingly active in using Nigeria as a base for the production of pirated goods.

Patent and trademark enforcement remains weak, and judicial procedures are slow and reportedly subject to corruption. However, the Nigerian government is taking steps to improve enforcement. Efforts to combat the sale of counterfeit pharmaceuticals, for example, have yielded some results. The Nigerian government also included pirated materials in the list of prohibited imports in the 2008-2012 CET Book, which provides NCS the authority to seize pirated works if imported into the country. In addition, theNigerian government has requested training to improve its enforcement efforts. In 2008, the UnitedStates responded by providing training assistance on IPR enforcement to Nigerian government officials in several sectors, with a focus on copyright piracy. The U.S. intends to continue IPR training assistance toNigeria in 2009, focusing on customs enforcement and combating counterfeiting that endangers public health and safety. Nigeria's broadcast regulations do not permit rebroadcast or excerpting of foreign programs unless the station has an affiliate relationship with a foreign broadcaster. This regulation is generally complied with, but some cable providers illegally transmit foreign programs. The National Broadcasting Commission monitors the industry and is responsible for punishing infractions. Widespread pirating of foreign and domestic videotapes discourages the entry of licensed distributors. In 2004, the Nigerian Copyright Commission (NCC) launched an anti-piracy initiative named "Strategy against Piracy." The Nigerian police force, working closely with the NCC, has raided enterprises producing and selling various pirated works such as software, books, and videos. About 29 cases are currently being prosecuted against IPR violators in various courts in the country. The Nigerian Economic and Financial Crimes Commission have also been active in IPR enforcement.

Foreign energy services suppliers are confronted with a number of barriers in Nigeria, particularly with respect to movement of personnel. Nigeria imposes quotas on foreign personnel based on the issued capital of firms. Such quotas are especially strict in the oil and gas sector and may apply to both production and services companies. Oil and gas companies must hire Nigerian workers unless they can demonstrate that particular positions require expertise not found in the local workforce. Positions in finance and human resources are almost exclusively reserved for Nigerians. Certain geosciences and management positions may be filled by foreign workers with the approval of the National Petroleum Investment and Management Services (NAPIMS) agency. Each oil company must negotiate its foreign worker allotment with NAPIMS. Significant delays in this process and in the approval of visas for foreign personnel present serious challenges to the energy industry in acquiring the necessary personnel for their operations. Investment in the petroleum sector is limited to existing joint ventures or production-sharing agreements. Foreign investors may buy shares of any Nigerian firm except those firms on an exemption list, which includes, for example, companies that manufacture firearms, ammunition, and military and paramilitary.

16.2.3. Foreign Trade Barriers

Foreign investors must register with the Nigerian Investment Promotion Commission after incorporation. Potential investors must contend with complex tax administration procedures, confusing land ownership laws, arbitrary application of regulations, corruption, and crime. The sanctity of contracts is often violated and Nigeria's court system for settling commercial disputes is weak and can be biased. Foreign oil companies are under significant pressure to increase procurement from domestic firms. The Nigerian government, through the Nigerian Content Division (NCD) of the Nigerian National Petroleum Corporation (NNPC), has set a target of 45 per cent local content for oil-related projects by 2006 and 70 per cent by 2010. The Nigerian government did not meet its target for 2008 due to infrastructure challenges such as electric power, and insecurity in the oil producing Niger Delta. In many cases, sufficiently trained personnel and physical infrastructure do not currently exist to meet the government's local content targets. Although some domestic firms possess adequate technical expertise, managerial and financial capabilities are often lacking. New legislation to codify mandatory levels of Nigerian content in specific petroleum activities is pending in the National Assembly. If enacted, the legislation would have a strong negative impact on the operations of foreign energy services companies already operating in Nigeria and could lead to a reduction in oil production. The vast majority of natural gas flaring in Nigeria is done in older, onshore, and near offshore oilfields.

International oil companies typically operate those fields in a joint venture arrangement with the state oil company as the majority partner. Funding for joint venture operations, maintenance, and equipment upgrades comes from joint venture partners in proportion to their equity ownership. Over the past several years, the Nigerian government has failed to fully fund its share of the joint venture costs, reducing the ability of the operating partners to install new anti-flare technology in these older oilfields.

16.3. Other Barriers

The Nigerian government has made efforts to eliminate financial crimes such as money laundering and advance fee fraud (also known as 419 frauds," after the relevant section of the Nigerian Criminal Code). In June 2006, the Financial Action Task Force removed Nigeria's name from the list of non cooperating countries and territories in the fight against money laundering and other financial crimes. In May 2007, Nigeria was admitted into the Egmont Group of Financial Intelligence Units. International monitoring groups routinely rank Nigeria among the most corrupt countries in the world, with the latest Transparency International rating being 121, an improvement from 147 in 2007, out of 180 countries. Despite this trend, Nigeria's corruption levels remain high and its main anticorruption institution, the Economic and Financial Crimes Commission has faltered recently in its reputation and commitments on the issue. Some U.S. suppliers believe they lose sales when they refuse to engage in illicit or corrupt behavior. Other U.S. exporters say Nigerian businessmen and officials understand that U.S. firms must adhere to the U.S. Foreign Corrupt Practices Act, and they believe that the law's restrictions help minimize their exposure to corruption.

17

Youth Empowerment and Development Drive

17.1. Youth Empowerment and Development Drive: An Invaluable Step to Secure Nigeria's Future Prosperity

Recent statistics from the Ministry of Youth Development shows that 28.14million Nigerian youths, representing 42% of the youth population, are unemployed. The statistics also indicates that 80% of the unemployed youth have only secondary school certificates and below. Some graduates have taken-up jobs that demean their worth. A number have fallen into crimes and social vices. Others have embarked on visionless emigration in search for 'greener pastures'. This appears irreconcilable with the fact that Nigeria has several growth reserves, which when spurred, could catapult the growth rate to 10% or more for a sustained period as China has experienced in the last two decades. For example, a substantial part of Nigeria's productive resources remain idle – waiting to be mobilized. Only 40% of the arable land is under cultivation, with 60% lying fallow. Much of the natural resources – oil, gas, bitumen, limestone, iron ore, columbite, gold, coal, gypsum, etc. – remain untapped.

Nigeria has the 6th largest deposit of natural gas in the world, and in the next few years, income from gas will surpass earnings from oil. Also, Nigeria possesses a reserve of resourceful population which provides the potential of a sound bridge of the future. Our large and expanding youthful population could provide the continuing growth dynamite as oppose to the ageing Western population. Then, what is the problem? No youth empowerment and development drive! According to Vivrus and Fletcher (2006), "Youth empowerment is an attitudinal, structural, and cultural process whereby young people gain the ability, authority, and agency to make decisions and implement change in their own lives and the lives of other people, including youths and adults." Meanwhile, Pittman (1993) defined youth development as "the ongoing growth process in which all youth are engaged in attempting to meet their basic personal and social needs to be safe, cared for, valued, useful, and spiritually grounded; and to build needed skills and competencies that allow them to function and contribute in their daily lives." [Sic]

A youth empowerment and development drive should take economic dimension, political dimension, and social dimension. The economic dimension should enable the youths to satisfactorily answer the question of how to meet basic human needs. Political dimension should give the youths a decisive say in governance. Social dimension will give the youths the platform to enjoy and protect social amenities provided and maintained by the government to enhance their physical and mental comfort. All these three areas of power interests are inseparable triplets. The presence of one is the security of the others. It is when the foregoing are achieved that the constitutional position that 'the welfare and the security of the people shall be the primary aim of government,' makes any sense in a modern society. In light of this, the rest of this essay will be expounded via three sections: Section 2.0, Section 3.0 and Section 4.0. Section 2.0 is the economic perspective which brings the plights of youths Not in Education, Employment or Training (NEET) into consideration and how it can be resolved. This section also considers the importance of the civic virtues of volunteerism, patriotism, and philanthropy in nation-building and economic development. Section 3.0 is the political perspective which seeks youth involvement in national decision making processvia youth-adult partnership; thereby proposing the establishment of a Youth Advocacy Council. Section 4.0 decries the erosion of morality, ethics and civic values and expresses the need for value orientation.

The large number of young people currently Not in Education, Employment or Training (NEET) is a significant challenge, not only from a macro perspective, but also in terms of its impact upon individual life chances. The Ministry of Youth Development should establish a Department of Business, Innovation and Skill so as to help develop a better understanding of the motivations and barriers affecting different groups of young people, with a view to offering better access to education and employment. More importantly, establishing this department should be with the sole aim of enhancing entrepreneurship. According to Adegbite (1992), "The current abysmal state of the economy as evidenced by rising unemployment calls for a re-appraisal of existing macro-economic frame. Thus, one of the most solid pillars for building a sound economy is having a solid SMEs sub-sector. In other words, for Nigeria to ensure a retreat from the state of economic stagnation and a return to the path of economic growth and development, she must find ways and means of nurturing a solid SME subsector." This is the whole essence of the proposed Department of Business, Innovation & Skills. As such, while a significant body of research already exists around the issues facing young people not in education, employment or training, the value of using young people's voice in improve the understanding of their aspirations and their attempts to realize them, is a key underpinning principle to finding appropriate ways to empower them.

17.2. Volunteerism, Patriotism, Philanthropy and Nation-Building

Though volunteering is not a common practice in Nigeria, it is noteworthy to emphasize that a few organizations do offer such opportunities to young people to be developed and contribute to the social economic growth of the country. Unfortunately, there is no strategic thrust for this; and because of the apparent lack of intention to develop this aspect, much of the work and processes are not documented. Volunteering would, ideally, reflect as a percentage of the nation's GDP. While definitions vary slightly, any definition of volunteering needs to contain three essential elements. Volunteering is done by choice, without monetary reward, and for the benefit of the community. 'Without monetary reward' does not exclude the payment of out-of-pocket expenses, which are a reimbursement for actual costs incurred rather than a reward.

There is a critical need for a volunteer and service movement in which the spirits of nation-building, patriotism and philanthropy will be engendered and perpetuated, allowing them to permeate our social fabric. A good way to go is for young people to begin to apply to organizations for voluntary work and involve themselves more in the social sector where the effects and impacts of their work are easily seen and felt. By doing these, the challenges and opportunities within the Nigerian context will be thrown up and formalized overtime. The bottom line is to build an effective and sustainable volunteer structure that can be easily reported and documented. To closely examine the relevance of the concepts of nation-building, patriotism and philanthropy, their interrelationships and impact on volunteering, a brief and basic description of each is useful here.

It is a general belief that nation-building is evolutionary rather than revolutionary; that it takes a long time and is a social process that cannot be jump-started from outside. Nation-building generally assumes that someone or something is doing the building intentionally and internally. Patriotism is reflected in the sum of loyal acts performed by the citizens of a nation towards that nation. The relevance of patriotism to the workings of volunteering cannot be over-emphasized. It is only when Nigerians, and indeed the youth population, feel a very deep and strong sense of patriotism that they can willingly volunteer their services, time and other resources. It is thus safe to say that volunteering is that link to building the nation intentionally as a social process, aims of which are achieved overtime with consistent and strategic efforts. And Nigerian youth can use volunteering as a springboard for this process, where they give back to the same society that birthed and nurtured them. Rather than focus on financial reward, young people should pursue the higher goal of emotional reward, intent on building their skills, gaining experience and contributing to the common good.

The economic miracle of Singapore under the leadership of Lee Kwan Yew epitomized this. Lee, in his best-seller book, "From THIRD WORLD to FIRST...The SINGAPORE STORY: 1956 – 2000", detailed the extra-ordinary efforts it took for an island city-state in Southeast Asia to survive at that time as a fledging nation-state. The arduous process of nation building was beaten small by creating an army of volunteers from a hitherto racially and ideologically divided population, thereby engendering national cohesion and integration. The government and philanthropists provided mass public housing, equipped schools with one computer for every two students, and connected every home to a broadband network. These were possible because Lee was a man of high integrity. This therefore calls for a shift in our artificial value system that is focused on money and what it can get for us so that we work to build our society rather than build our individual store houses. This is a way of reducing social ills and ensuring that the larger and more energetic part of society can work to uplift the less privileged. Volunteerism can only thrive where citizens are patriotic and committed to building their society. It is not rocket science. It takes a lot of courage, purposefulness and resilience, and all these most youth possess in abundance. The following statement is credited to former Apartheid South African leader, P.W. Botha,

> *"...Pretoria has been made by the White mind...*
>
> *The Republic of South Africa that we know of today has not been created by wishful thinking. We have created it at the expense of intelligence, sweat and blood.*
>
> *...By now every one of us have seen it practically that the Blacks cannot rule themselves. Give them guns and they will kill each other...let us all accept that the Black man is the symbol of poverty, mental inferiority, laziness and emotional incompetence.*
>
> *...Blacks are vulnerable to money inducements. The old trick of divide and rule is still very valid today...His inferior sense of morals can be exploited beautifully. AND HERE IS A CREATURE THAT LACKS FORESIGHT" (Quadri, 2013).*

Driving this statement home to the Nigerian context, mindless of the fact that P.W. Botha is a cultural bigot and undeserving of being quoted, the veracity of parts of the write up can hardly be denied by honest and trueborn Nigerians. For instance, although Botha's position that blacks are incapable of leadership is totally unacceptable, the leadership trends in Nigeria have not disproved it. And then, alas, there is a subtle apartheid between the youth and the adults in ruling the affairs of this nation: the *'Nigerian neo-apartheid'* scenario. Adults think that leadership is all about age, and hence leave out youth involvement, forgetting that young minds are the think-tank of innovation and ingenuity. No wonder Nigeria is where she is today.

Not to sound pessimistic or like a prophet of doom, if the much contemplated Sovereign National Conference neglects youths of just reputation and intellectual disposition, we are heading nowhere. This can be told of Lee Kwan Yew's approach to transforming Singapore. His government provided scholarship to excellent youths to study in Harvard, Princeton, Cambridge, Oxford, Yale and the likes then return home to implement social change with the superiority of reasoning acquired from these institutes of higher learning. They were the first-class minds that surrounded him in taking decisions of national interest. Gone are the days when the involvement of youths was not considered significant in societal development. Today, it is a different case because elders and sages have come to realize that the youth of any nation are the emblem of its heritage. In fact, youth are the embodiment of a nation's historical past, present reality and future possibilities. A popular Yoruba maxim reads, "it is through the joint wisdom of both youths and adults that Ife (of ancient time) came into existence"; it is through the joint efforts of the youths and adults of Nigeria that the future prosperity of Nigeria can be secured.

Creating a Youth Advocacy Council will make a real difference for countless young people. It will empower teenagers and young adults caught up in government systems to take positive control of their lives through learning to advocate for themselves. The Youth Advocacy Council will also provide scholarship to exceptional students who are interested in *civic entrepreneurship* to study; after which they will be automatically employed into the council as field leaders to provide pragmatic initiatives that will push the government to create new policies, practices, and programmes. The Youth Advocacy Council is to set up model self-advocacy programs at law schools, to inspire law students to embrace working with *at-risk youth* (that is, youths with dashed hopes for the future, youths with a history of trouble with the law, teenage mothers, etc). This will make for due consideration of the otherwise ignored youth.

The Youth Advocacy Council should be the voice calling for attention to be paid to the particular needs and strengths of the young people who are forgotten, overlooked and failed by bureaucracies; thus, arousing a deepened and more sophisticated national and local understanding of issues facing youths. This will enable the enactment of better laws and policies that address youth's needs and attract organizations and funders to turn their attention towards youth development. A Youth Advocacy Council is a sure way for all youth, especially those living in poverty, to achieve equal opportunities and social justice. This is a necessary condition to ensure a sense of hope and confidence in the resilience and ability of young people to become successful participating citizens in our community.

The issues of morality, ethics and values are critical to sustainable development of Nigeria. These concepts have been used by different people to mean different things, and meanings deployed depend on the cultural background of the user and the context of use. Generally however, morality, ethics or values could be collectively described as those standards of conduct that apply to everyone in the society rather than only to members of a special group. It is expected that these standards of conduct are those that every rational person in the society wants every other person to comply with. For instance, life will be lot easy if everyone queues up at the Bus-Stop, Bank/ATM, etc. In the Nigeria of 1960s and early 1970s when corruption was less pervasive in society as it is today, young Nigerians learnt from home such basic moral rules as: don't lie; don't kill; don't cheat; keep your promises; don't steal; and so on. These moral lessons were usually based on religious tenets, community mores or a combination of both. This means that by the time students were ready to enroll in the university they were morally oriented to face the challenges in the society. Today, I do not think we can say the right level of moral preparation of young Nigerians exists in the society. In fact, some Nigerian youths have become skilled in corrupt practices by emulating the older generation, for instance, in cheating during examinations. This has further accentuated as the decay in the social values continued by the years with the attendant repercussion in terms of financial, economic, educational and political corruption. In the 1960s and early 1970s, the incidents of and frequency of public discourse on corruption were minimal. Though corrupt practices existed in these early times of the Nigerian nation, the current dimensions have far exceeded what they used to be. The financial scandals in 2012 alone include pension funds (billions of naira), fuel subsidy (trillions of naira) and sale of OPL245 oil fields. Corruption is now the top priority development challenge facing the three tiers of government in Nigeria. As such, there is no doubt that all these have resulted to deep crass for materialism and poor political values (the 'share-the-national-cake-syndrome') among some aspiring Nigerian youths who want to get rich quickly by occupying public offices. Indeed, there is need for value re-orientation! The requisite values include exemplary leadership, integrity, concern for the welfare of the people, objectivity and truthfulness, faithfulness and reliability/dependability. The opposite of these behaviours must be avoided.

Let no Nigerians claim that we have not burdened ourselves in our ambitious, but legitimate, effort to transform our society and secure our future national prosperity within these fifty three years of independence; however, youth empowerment and development is the missing link. I feel compelled to agree to the subjective fact that it is more difficult to manage affluence than to manage poverty. However, we must drive our way through to affluence. For a nation with such a rich diversity of youth, driving our way through to affluence requires youth development in all shapes and sizes. To achieve this, the following are needed:

1. A leadership development programme that offers youth neutral territory where they can build skills and relate to one another as individuals;
2. A city government that engages the youth in the policy making process through youth councils and youth positions in government departments;
3. A religious institution that provides the youth with proper moral training; and
4. A local business which employs youth in meaningful and relevant work.

17.3. Foreign Direct Investment (FDI)

Foreign direct investment (FDI) plays an extraordinary and growing role in global business. It can provide a firm with new markets and marketing channels, cheaper production facilities, access to new technology, products, skills and financing. For a host country or the foreign firm which receives the investment, it can provide a source of new technologies, capital, processes, products, organizational technologies and management skills, and as such can provide a strong impetus to economic development. Foreign direct investment, in its classic definition, is defined as a company from one country making a physical investment into building a factory in another country. The direct investment in buildings, machinery and equipment is in contrast with making a portfolio investment, which is considered an indirect investment. In recent years, given rapid growth and change in global investment patterns, the definition has been broadened to include the acquisition of a lasting management interest in a company or enterprise outside the investing firm's home country. As such, it may take many forms, such as a direct acquisition of a foreign firm, construction of a facility, or investment in a joint venture or strategic alliance with a local firm with attendant input of technology, licensing of intellectual property, In the past decade, FDI has come to play a major role in the internationalization of business. Reacting to changes in technology, growing liberalization of the national regulatory framework governing investment in enterprises, and changes in capital markets profound changes have occurred in the size, scope and methods of FDI. New information technology systems, decline in global communication costs have made management of foreign investments far easier than in the past. The sea change in trade and investment policies and the regulatory environment globally in the past decade, including trade policy and tariff liberalization, easing of restrictions on foreign investment and acquisition in many nations, and the deregulation and privatization of many industries, has probably been the most significant catalyst for FDI's expanded role.

The most profound effect has been seen in developing countries, where yearly foreign direct investment flows have increased from an average of less than $10 billion in the 1970's to a yearly average of less than $20 billion in the 1980's, to explode in the 1990s from $26.7billion in 1990 to $179 billion in 1998 and $208 billion in 1999 and now comprise a large portion of global FDI. Driven by mergers and acquisitions and internationalization of production in a range of industries, FDI into developed countries last year rose to $636 billion, from $481 billion in 1998 (Source: UNCTAD) Proponents of foreign investment point out that the exchange of investment flows benefits both the home country (the country from which the investment originates) and the host country (the destination of the investment). Opponents of FDI note that multinational conglomerates are able to wield great power over smaller and weaker economies and can drive out much local competition. The truth lies somewhere in the middle.

For small and medium sized companies, FDI represents an opportunity to become more actively involved in international business activities. In the past 15 years, the classic definition of FDI as noted above has changed considerably. This notion of a change in the classic definition, however, must be kept in the proper context. Very clearly, over 2/3 of direct foreign investment is still made in the form of fixtures, machinery, equipment and buildings. Moreover, larger multinational corporations and conglomerates still make the overwhelming percentage of FDI. But, with the advent of the Internet, the increasing role of technology, loosening of direct investment restrictions in many markets and decreasing communication costs means that newer, non-traditional forms of investment will play an important role in the future. Many governments, especially in industrialized and developed nations, pay very close attention to foreign direct investment because the investment flows into and out of their economies can and does have a significant impact. In the United States, the Bureau of Economic Analysis, a section of the U.S. Department of Commerce, is responsible for collecting economic data about the economy including information about foreign direct investment flows. Monitoring this data is very helpful in trying to determine the impact of such investments on the overall economy, but is especially helpful in evaluating industry segments. State and local governments watch closely because they want to track their foreign investment attraction programmes for successful outcomes.

As mentioned above, the overwhelming majority of foreign direct investment is made in the form of fixtures, machinery, equipment and buildings. This investment is achieved or accomplished mostly via mergers & acquisitions. In the case of traditional manufacturing, this has been the primary mechanism for investment and it has been heretofore very efficient. Within the past decade, however, there has been a dramatic increase in the number of technology start-ups and this, together with the rise in prominence of Internet usage, has fostered increasing changes in foreign investment patterns. Many of these high tech start ups are very small companies that have grown out of research & development projects often affiliated with major universities and with some government sponsorship. Unlike traditional manufacturers, many of these companies do not require huge manufacturing plants and immense warehouses to store inventory. Another factor to consider is the number of companies whose primary product is an intellectual property right such as a software program or a software-based technology or process. Companies such as these can be housed almost anywhere and therefore making a capital investment in them does not require huge outlays for fixtures, machinery and plants.

In many cases, large companies still play a dominant role in investment activities in small, high tech oriented companies. However, unlike in the past, these larger companies are not necessarily acquiring smaller companies outright. There are several reasons for this, but the most important one is most likely the risk associated with such high tech ventures. In the case of mature industries, the products are well defined. The manufacturer usually wants to get closer to its foreign market or wants to circumvent some trade barrier by making a direct foreign investment. The major risk here is that you do not sell enough of the product that you manufactured. However, you have added additional capacity and in the case of multinational corporations this capacity can be used in a variety of ways. High tech ventures tend to have longer incubation periods. That is, the product tends to require significant development time. In the case of software and other intellectual property type products, the product is constantly changing even before it hits the marketplace. This makes the investment decision more complicated. When you invest in fixtures and machinery, you know what the real and book value of your investment will be. When you invest in a high tech venture, there is always an element of uncertainty. Unfortunately, the recent spate of dot.com failures is quite illustrative of this point.

Therefore, the expanded role of technology and intellectual property has changed the foreign direct investment playing field. Companies are still motivated to make foreign investments, but because of the vagaries of technology investments, they are now finding new vehicles to accomplish their goals. Consider the following:

- ☆ ***Licensing and technology transfer***. Licensing and tech transfer have been essential in promoting collaboration between the academic and business communities. Ever since legal hurdles were removed that allowed universities to hold title to research and development done in their labs, licensing agreements have helped turned raw technology into finished products that are viable in competitive market places. With some help from a variety of government agencies in the form of grants for R&D as well as other financial assistance for such things as incubator programs, once timid college researchers are now stepping out and becoming cutting edge entrepreneurs. These strategic alliances have had a serious impact in several high tech industries, including but not limited to: medical and agricultural biotechnology, computer software engineering, telecommunications, advanced materials processing, ceramics, thin materials processing, photonics, digital multimedia production and publishing, optics and imaging and robotics and automation. Industry clusters are now growing up around the university labs where their derivative technologies were first discovered and nurtured. Licensing agreements allow companies to take full advantage of new and exciting technologies while limiting their overall risk to royalty payments until a particular technology is fully developed and thus ready to put new products into the manufacturing pipeline.

☆ ***Reciprocal distribution agreements.*** Actually, this type of strategic alliance is more trade-based, but in a very real sense it does in fact represent a type of direct investment. Basically, two companies, usually within the same or affiliated industries, agree to act as a national distributor for each other's products. The classical example is to be found in the furniture industry. A U.S.-based manufacturer of tables signs a reciprocal distribution agreement with a Spanish-based manufacturer of chairs. Both companies gain direct access to the other's distribution network without having to pay distributor support payments and other related expenses found within the distribution channel and neither company can hurt the other's market for its products. Without such an agreement in place, the Spanish manufacturer might very well have to invest in a national sales office to coordinate its distributor network, manage warehousing, inventory and shipping as well as to handle administrative tasks such as accounting, public relations and advertising.

☆ ***Joint venture and other hybrid strategic alliances.*** The more traditional joint venture is bi-lateral, that is it involves two parties who are within the same industry who are partnering for some strategic advantage. Typical reasons might include a need for access to proprietary technology that might tip the competitive edge in another competitor's favour, desire to gain access to intellectual capital in the form of ultra-expensive human resources, access to heretofore closed channels of distribution in key regions of the world. One very good reason why many joint ventures only involve two parties is the difficulty in integrating different corporate cultures. With two domestic companies from the same country, it would still be very difficult. However, with two companies from different cultures, it is almost impossible at times. This is probably why pure joint ventures have a fairly high failure rate only five years after inception. Joint ventures involving three or more parties are usually called syndicates and are most often formed for specific projects such as large construction or public works projects that might involve a wide variety of expertise and resources for successful completion. In some cases, syndicates are actually easier to manage because the project itself sets certain limits on each party and close cooperation is not always a prerequisite for ultimate success of the endeavour.

- ***Portfolio investment.*** Yes, we know that you're paying attention and no we're not trying to trip you up here. Remember our definition of foreign direct investment as it pertains to controlling interest. For most of the latter part of the 20th century when FDI became an issue, a company's portfolio investments were not considered a direct investment if the amount of stock and/or capital was not enough to garner a significant voting interest amongst shareholders or owners. However, two or three companies with "soft" investments in another company could find some mutual interests and use their shareholder power effectively for management control. This is another form of strategic alliance, sometimes called "shadow alliances". So, while most company portfolio investments do not strictly qualify as a direct foreign investment, there are instances within a certain context that they are in fact a real direct investment.

Direct foreign investment allows companies to accomplish several tasks:

- Avoiding foreign government pressure for local production.
- Circumventing trade barriers, hidden and otherwise.
- Making the move from domestic export sales to a locally-based national sales office.
- Capability to increase total production capacity.
- Opportunities for co-production, joint ventures with local partners, joint marketing arrangements, licensing, etc.

A more complete response might address the issue of global business partnering in very general terms. While it is nice that many business writers like the expression, "think globally, act locally", this often used cliché does not really mean very much to the average business executive in a small and medium sized company. The phrase does have significant connotations for multinational corporations. But for executives in SME's, it is still just another buzzword. The simple explanation for this is the difference in perspective between executives of multinational corporations and small and medium sized companies. Multinational corporations are almost always concerned with worldwide manufacturing capacity and proximity to major markets. Small and medium sized companies tend to be more concerned with selling their products in overseas markets. The advent of the Internet has ushered in a new and very different mindset that tends to focus more on access issues. SME's in particular are now focusing on access to markets, access to expertise and most of all access to technology.

Depending on the industry sector and type of business, a foreign direct investment may be an attractive and viable option. With rapid globalization of many industries and vertical integration rapidly taking place on a global level, at a minimum a firm needs to keep abreast of global trends in their industry. From a competitive standpoint, it is important to be aware of whether a company's competitors are expanding into a foreign market and how they are doing that. At the same time, it also becomes important to monitor how globalization is affecting domestic clients. Often, it becomes imperative to follow the expansion of key clients overseas if an active business relationship is to be maintained. New market access is also another major reason to invest in a foreign country. At some stage, export of product or service reaches a critical mass of amount and cost where foreign production or location begins to be more cost effective. Any decision on investing is thus a combination of a number of key factors including:

- ☆ Assessment of internal resources,
- ☆ Competitiveness,
- ☆ Market analysis,
- ☆ Market expectations.

From an internal resources standpoint, does the firm have senior management support for the investment and the internal management and system capabilities to support the set up time as well as ongoing management of a foreign subsidiary? Has the company conducted extensive market research involving both the industry, product and local regulations governing foreign investment which will set the broad market parameters for any investment decision? Is there a realistic assessment in place of what resource utilization the investment will entail? Has information on local industry and foreign investment regulations, incentives, profit retention, financing, distribution, and other factors been completely analyzed to determine the most viable vehicle for entering the market (green field, acquisition, merger, joint venture, etc.

18
Global Sustainability

18.1. Global Sustainability

The quest for global sustainable development is espoused by a multitude of governments, intergovernmental organizations, non-governmental organizations and, most recently, private companies. The World Summit on Sustainable Development (WSSD), held in Johannesburg in 2002, was the latest global forum to assess progress in achieving sustainable development and to set priority areas for continued efforts, but international political and scientific efforts on the simultaneous achievement of environmental protection and social and economic development started long before WSSD. Four decades of extensive high-level international cooperation and policy making on environment and development, however, have proven to be a long and difficult road towards a global environmentally sound development, mixing a few policy successes with a frequent lack of effective implementation and behavioral changes.

The UN Charter from 1945 assigns governments a responsibility to work for peace, freedom, human rights and social and economic progress for all people. Yet, the Charter was silent on the role of environmental issues in achieving these goals. As indications of widespread ecological deterioration became acknowledged in the 1960s, voices were quickly raised for expanding UN activities into the environmental field. In 1970, UN Secretary-General U Thant remarked that "never in the twenty-fiveyear history of the United Nations has there been a problem of more relevance to all nations than the present environment crisis," and added that "the time has come for the United Nations, in the spirit and letter of the Charter, to be a center for harmonizing the actions of nations in solving the problems of the human environment" (Cordier and Harrelson, 1977: 346). Promoting global environmentally sound social and economic development has been a central policy challenge for the United Nations and international and domestic politicians and policy makers since the 1960s. In an ongoing, multi-year research project, we study global attempts under to integrate environment and development issues on conceptual and institutional levels, as well as efforts to translate these into international and domestic action. These issues are not only of historical interest, but they can give us a better understanding of the origin of the ongoing sustainability debate and highlight some of the central factors that continue to influence cooperation and policy making on sustainable development. We argue that some of the major obstacles towards more effective implementation and progress can be found in the very factors that influence and shape the policy process.

In this book, we look at major political events, efforts and policy developments on environment and development that preceded the World Commission on Environment and Development (the so-called Brundtland Commission). We argue that these form an important conceptual, political and institutional background to the Brundtland report *Our Common Future*, published in 1987, which popularized the concept of sustainable development and brought it into the center of the environment and development debate. The paper identifies factors influencing cooperation and policy making on environment and development and discusses how these explanatory factors shape policies and outcomes (or lack thereof). We track these issues up to the United Nations Conference on Environment and Development (UNCED), held in Rio de Janeiro, in 1992. Much of the existing academic literature on international cooperation and policy making on sustainable development looks at the sustainable development issue from a comparatively short historical perspective (Toman, 1994; Goodland, 1995; Bartelmus, 2003; Howarth, 1997; Faucheux, Pearce and Proops, 1996). This literature generally begins in the early 1980s with the creation of the Brundtland Commission and views *Our Common Future* as one of the first substantial international policy attempts on environment and development. As a result, there is little in-depth literature on where the Brundtland report came from – was it a completely new look at the problems of environment and development, or can it better be explained as a continuation of a longer process of debate and cooperation on these issues?

We believe that a much more detailed examination of pre-Brundtland events on environment and development will help us better understand the conceptual and political background of the Brundtland Commission, its report and subsequent international political and policy efforts on sustainable development. A deeper understanding of how the environment and development issue was shaped in the past will give us a better understanding of ongoing cooperation and policy making on issues relating to sustainable development. As a result, this discussion paper emphasizes and examines the process of identifying those central factors that were most crucial in shaping the global environment and development agenda up to UNCED, and the political efforts to address these issues during this time. In this book,we apply four analytical perspectives. These analytical perspectives, which all exist in the international relations literature but have not been systematically applied and compared to each other in the area of environment and development, are: the emergence and influence of an international environment and development discourse; an international political push for more multilateralism and building of new multilateral institutions; power politics including the influence of Cold War relations; and North-South politics and conflicts. We use these analytical perspectives to seek a deeper understanding of which forces were most important in shaping early international cooperation and policy making on environment and development, and how this cooperation and policy making fed into the sustainability debate and policy process that began in the late 1980s, and still continues.

Example is the preparations for, holding of and outcomes of the 1972 United Nations Conference on the Human Environment (UNCHE), which was the first global political conference that was designed to simultaneously address environment and development issues. In this section, we look at the conceptualization and debate of a threatening ecological crisis shaping the conference agenda and discussions, characters and implications of East-West and North-South dimensions of international politics and conflict on environment and development issues, and calls for increased multilateral cooperative efforts and attempts at institution building on environment and development issues. Also, we look at international debate and policy actions on environment and development issues in the early 1970s to early 1980s, including the holding of the Cocoyoc Symposium in 1974 and the launching of the World Conservation Strategy in 1980 and the World Charter for Nature in 1982. Here, we focus on central themes of the environment and development debate, accelerated attempts at international ecological preservation through the creation of multilateral organizations, agreements and programs, the role of international organizations and countries pressing for multilateral action, the continuation of EastWest mistrust and Cold War politics, and how persisting North-South conflicts influence political and policy efforts on environment and development issues.

18.1.1. Emergence and Influence of an International Environment and Development Discourse

One possible explanatory factor relating to the global politics and policy making on environment and development is the notion of a severe environmental crisis due to uncontrolled ecological deterioration, triggering demands for policy responses. Ecological deterioration and loss of species received international political attention already in the early 20th century, and political, media and public attention to issues of pollution, resource depletion and population grew after WWII. Scholars have demonstrated the existence of an environmental discourse by the 1960s, particularly in industrialized societies (Hajer, 1995; McNeill, 2000; Worster, 1994; Dalton, 1994; Hays, 1985; Opie, 1998; Linnér, 2003). The term "discourse" in this sense characterizes an ongoing historic process of creating meaning by systematizing statements. A discourse is expressed by talking about a particular object, as well as through its related symbols, practices and institutions. Symbols should be understood in relation to their specific use within a discourse, and the rules that guide this use; what the French philosopher Michel Foucault calls their 'grammatics.' These selective mechanisms involve claims of truth and rules of what it is possible and not possible to say, and who is allowed to speak (Foucault, 1972: 219-225). This discourse can be identified and analyzed by examining, for example, relevant political statements and speeches, policy debates and policy documents, popular and academic studies, and public perceptions.

By the 1960s 'the environment' was formed into a distinct international and domestic political field with its own issues, symbols, practices and institutions. The growing political belief that accelerating industrialization could result in a severe international ecological crisis and the resultant discursive position of environmental issues explains, at least in part, why the environment became a priority issue area in international politics in the 1960s. In addition to concerns that were expressed by international and domestic policy makers, there was a growing public pressure on policy makers in many industrialized countries to deal with both domestic and transboundary environmental threats. The environmental discourse early on connected issues of ecological deterioration and the growing demand for natural resources with human development issues. As such, the international environmental debate and discourse from the beginning included and addressed intersecting environment and development issues (Bäckstrand, 1971; Rambach, 1971; Campbell, 1973; Engfeldt, 1973). Issues that were raised for industrialized countries included levels of consumption, resource use, and air and water pollution and its health effects (Carson, 1962; Rowland, 1973; Caldwell, 1996). Concerns for developing countries focused on population issues and the improvement of social conditions (for example, housing and access to clean water) and economic conditions (for example, financial wealth) (Ehrlich, 1968; Meadows *et al.*, 1972; Rowland, 1973). This environment and development related international discourse has evolved over the past four decades, up until the present time, shaping perspectives and debates and feeding into cooperation and policy making. In this paper, we identify major components of this environment and development discourse at our four historical periods and link these to related cooperation and policy developments.

18.1.2. Growing Multialteralism and the Influence of International Institutions

Scholars have demonstrated how a dramatic increase in the number of international organizations over the past century has facilitated a never before seen level of multilateral cooperation and activities, linking countries and people (Beck 2000; Jameson 1998a, 1998b; Routledge, 2002). An increase in transnational transportation and flows of capital, resources, materials, substances, communication infrastructure and mental globalization may reinforce this interdependence between countries and people (Keohane and Nye, 2001; Castells, 1996, 1997, 1998). International institutions and transnational flows of capital, resources, and information delineate common standards for behavior that reach and can affect abilities and choices by local actors. In turn, a multitude of local actions such as the election of political leaders and consumer decisions influence the behavior and decisions by transnational actors (Jasanoff and Long Martello, 2004).

In this book, we discuss how efforts to promote and design multilateral cooperation on environment and development issues have occurred over time. In addition, we examine the influence of international institutions on cooperation and policy making. Many countries, particularly smaller non-aligned ones, saw UN involvement in environment and development issues as a way to promote both international cooperation on these issues and strengthen the role of the United Nations through the formulation of related principles, agreements, organizations and programs (Engfeldt, 1973; UNCHE Country Reports). These strives toward increased multilateralism and UN involvement, however, were opposed by other countries, protective of their national sovereignty. This resistance has been expressed both by larger industrialized countries and developing countries. These conflicts were visible in, for example, the development of international environmental law and the creation of the United Nations Environment Programme (Campbell, 1973; Rowland, 1973; Caldwell, 1996; Strong, 2001). In addition, in a context of a growing number of international organizations and policy efforts addressing environment and development issues, representatives of international organizations often argue for their involvement in international policy making and implementation. In our case study on international cooperation and policy making on environment and development, we see evidence of this. As many UN specialized agencies and other international organizations such as the International Union for the Protection of Nature (IUPN, from 1953 the International Union for the Conservation of Natural Resources, IUCN) and the International Council of Scientific Unions (ICSU) became involved in the environment and development agenda, they often supported calls for more international cooperation. They also sought to keep specific environment and development issues on the international agenda through organizing meetings, creating programmes and publishing reports.

International relations literature frequently looks at traditional power politics to explain international politics and outcomes. From a Realist perspective focusing on the role of material capabilities, international politics and outcomes are seen as resulting from the cost-benefit assessments of advantages made by the most powerful state(s). International politics is seen as a zero-sum game: one state's gain is inevitable another state's loss. Calculations by states are based on (but not exclusively determined by) the preferences (utility functions) of those states. This view is reflected in the perhaps most well known power-based explanation, Robert Keohane's hegemonic stability hypothesis (Keohane, 1980). Expressing a general skepticism about the possibility of international cooperation, the hegemonic stability hypothesis argues that institutions will emerge only as the result of independent action by one supreme state that holds the necessary power resources.

Undoubtedly, power calculations and exercise of material resources related to national security can be important in shaping international politics and outcomes. In this paper, we explore cases where power and domestic security consideration influence cooperation and policy making on environment and development related issues. For example, international action on species protection in the early 1900s included European powers imposing species protection measures in their foreign colonies and territories. Between 1945 and the early 1990s, Cold War politics and the power struggle between the United States and the Soviet Union were dominating factors for much international cooperation and policy making. Cold War politics and East and West competition and cooperation also influenced international policy making on environment and development issues. The bi-polar international political order that emerged shortly after WWII included strong geopolitical security concerns and mutual efforts by the United States and the Soviet Union to balance their power and political influence in the international political system. These geopolitical security concerns contributed to a political and scientific focus on natural resource security as a factor of social and political stability and instability in individual countries and the international system. This is evident in, for example, issues relating to the geopolitics and science of natural resource management in the 1940s to 1960s (Linnér, 2003) and the preparations for and holding of the United Nations Conference on the Human Environment in 1972 (Rowland, 1973; Strong, 2001). In addition, the environment was used as an issue area for building trust between the two blocs during times of détente. This was the case in the late 1960s as "the environment" became an international political issue (Caldwell, 1996), and can also be seen in efforts to tackle transboundary emissions of sulfur and other air pollutants under the United Nations Economic Commission for Europe during the 1970s (Chossudovsky, 1989).

18.2. North-South Politics and Conflicts

International relations literature has looked at North-South politics and issues. Beginning at the Bandung conference in Java in 1955, many developing countries took on a more independent role in international politics. At the conference, 29 Asian and African countries, among others Burma, the People's Republic of China, Ceylon, the Indian Union and Indonesia proclaimed themselves to be a third force, beside the Eastern and Western blocs. During the period from 1955 to 1971 UN membership grew rapidly, as former colonies became independent nations. The organization had started with 51 members in 1945. By 1959 total membership had reached 83. By 1972 there were 132 members, many of these former colonies (Goodrich, 1974). With these new nations, a growing number of UN member states had an interest in strengthening the role of the United Nations Economic and Social Council (ECOSOC) on behalf of the UN Security Council and the powerful members of the East and West blocs (Arrighi, 1994; Cammack *et al.*, 2001; Neuman, 1998; Keohane and Nye, 2001).

In this book, we examine the influence of North-South relations and issues on attempts to formulate and implement effective environment and development policy and action in both industrialized and developing countries. Many developing countries have argued that the accelerating globalization of production and consumption have led to an increased dependence of the powerless on the powerful. Repeatedly, they have claimed that the global economy conferred an uneven exchange upon developing and industrialized countries because of the structure of an economic and trade system that was built on the North's domination in five areas: technological distribution and the capacity of large-scale investments; control of worldwide financial markets, as finance capital was becoming an increasingly important component of capitalist economy; control over the planet's natural resources, either through multinational cooperation or dependency in developing countries on income from exporting non-refined resources; media and communication monopoly; and research and development capacity to produce new technology and substances, identify harmful effects and develop alternatives (Castells, 1996, 1997, 1998; Amin, 1997).

This dependence of the South on the North is not restricted to economic issues. Scholars have argued that the least developed countries have had limited national sovereignty and ability to act across a range of political issues since the colonial period (Neuman, 1998: 9-10). This critique argues that the international division of production causes a lack of effective control over economic resources in developing countries, which maintains their dependency on the global North. This lack of control over natural resources threatens to challenge the legitimacy of domestic political order (Inayatullah, 1996). Several researchers have argued that most conflicts in developing countries during the Cold War were caused by regional or domestic factors, such as weak postcolonial state structures and challenged legitimacy of rulers (Inayatullah, 1996; Ayoob, 1984: 477-93, 1992; David, 1991: 232-56; Acharya, 1998: 172-77). This may in turn partially explain the resolve among several political leaders in the South to challenge the international economic order through the industrialized countries' environmental concern.

The detrimental effects of global political and economic systems on the South were not only emphasized by politicians from the South, but were also highlighted by decision makers, media and the public in the North. Concern in the North was expressed both because of a growing opinion in the North that the North could not resolve global problems without active cooperation by the developing countries, and also as a sign of an emerging ethical consciousness that emphasized the global household and a moral obligation of people in the North towards people in the South. For example, Zbigniew Brzezinski (1973: 726) argued that the growing inequality between developing and developed nations was becoming "the most powerful moral imperative of our time, thus paralleling the appeal of the concept of liberty during the nineteenth century." Regardless of whether concerns over the North-South divide were expressed as a result of moral stance or of self interest, or a combination of the two, North-South relations became a crucial issue on the United Nations agenda in the 1960s.

Countries in the North and in the South often expressed different interests and preferences on environment and development issues. In this paper, we categorize the differing views between the North and the South on environment and development in terms of universalizing and differentiating views, building on earlier work by, among others, Beck (2000), Jameson (1998a; 1998b), Taylor and Buttel (1992) and Sachs (1993). These views will be discussed throughout this paper, but briefly stated, a universalizing perspective views the increasing importance of environment and development issues for accentuating common interests and the need for joint action. Based on such a view, many industrialized countries stressed that all countries and people are bound together by common fundamental interests on environment and development (UNCHE Country Reports). Early technical, scientific and political efforts on environment and development often became a vehicle for reinforcing such a view.

However, attempts to formulate international policy and programmes on environment and development issues were more complex than simply promoting common interests. Many practitioners and scholars have strongly criticized the universalizing view on environment and development in which solutions to environment and development issues are seen to be based on common interests among countries and people formulated by the North (Harvey, 1990; Taylor and Buttel, 1992; Shiva, 1992; Sachs, 1993; Hajer, 1995; Guha and Martinez-Alier, 1997; Jameson, 1998a, 1998b). Hajer and Fischer (1999: 2-5) have argued that much of the early international environment and development discourse suggested much more unity and shared understanding between the countries and the cultures of the world than can be legitimately assumed. Many politicians and experts from developing countries acknowledged an accentuation of world interdependence, but questioned central conceptual and material aspects of this process (Rambach, 1971; Rowland, 1973).

While many UN efforts and other international actions on environment and development were intended to stimulate cooperation and policy making on environment and development, they also provided an opportunity to criticize the notion of one world - one interest. From the 1960s and onwards, representatives of developing countries repeatedly criticized what they saw as efforts by industrialized countries to preserve the dominance of the global North (Rambach, 1971; Rowland, 1973). Instead, they called for the creation of institutions and practices that reflected differences in situations and interests between industrialized and developing countries. Such calls were, for example, connected to calls for a new international economic order that would be more geared towards the conditions and interests of developing countries (Furtado, 1971; Furtado, 1976; Cardoso, 1979).

The conflict between proponents of universalizing and differentiating views on environment and development also concerns which specific issues are prioritized for global action. Many industrialized countries argued for only minor changes in the international economic system, and instead stressed the need for concerted action on international environmental issues (Åström, 1972; Campbell, 1973; Engfeldt, 1973). It was also industrialized countries that pressed the hardest for international action on transboundary environmental issues at the Stockholm Conference and in the aftermath of Stockholm (UNCHE Country Reports; Caldwell, 1996). Such transboundary issues included marine and river pollution, air pollution, and the trade and dispersal of hazardous substances (Caldwell, 1996). In contrast, many developing countries have argued that primary focus should be on development issues (Rambach, 1971; UNCHE Country Reports). They have argued that economic and social development is the most effective way to address environmental concerns in developing countries (which is the majority of the world's population). Therefore, the international community should pay greater attention to improving social and economic conditions for the poorest in order to address both development and environment issues, according to many developing countries. Concerns expressed over environmental issues by developing countries often included local land use problems, improved access to safe drinking water, and issues relating to rapid urbanization (Rambach, 1971; UNCHE Country Reports).

18.3. The Discourse of Species Protection and Natural Resource Management

The international discourse of species protection at the beginning of the 20th century was highly westerncentered. It emerged out of domestic ecological protection efforts in Europe and North America that were expanded to include the organizing of international conferences and the negotiations of regional multilateral agreements. In some cases, it was aesthetic concerns of preserving "beautiful" wildlife that prompted such action (Hayden, 1942), but in others, countries acted to ensure long-term social and economic rewards from controlled harvesting (Barrett, 2003). In addition, northern countries debated the need to species protection efforts in foreign colonies and territories, and also imposed several sets of regulations in their colonies and territories up until WWII. After WWII, the discourse began to change. While multilateral efforts on species protection continued, countries began debating larger issues of natural resource use and management. This change in the international discourse was fuelled by scientific and technological development and geopolitical security concerns that were related to the Cold War and intensifying competition between East and West. This discourse moreover included debates of threatening conditions in developing countries. Fears were expressed that "underdevelopment" and instability in newly independent countries could result in violent conflict with potentially severe domestic and international implications (Caldwell, 1996: 51; Linnér, 2003: 61-92).

Issues relating to a rapid population growth became a central issue in an emerging environment and development discourse in the 1960s. Again, concerns were voiced that a "population bomb" in developing countries and related demands for natural resources could cause instability and conflict with dire local and international consequences (Ehrlich, 1968; Meadows *et al.*, 1972). At the same time, many larger developing countries began taking a much more independent stand in the United Nations in the 1960s, arguing that the international community had a responsibility to pay more attention to local and domestic development issues in the global South. In addition, public and political attention to pollution issues increased sharply in mainly industrialized countries in the 1960s (Carson, 1962), leading to calls for increased multilateral cooperation on transboundary pollution issues (Caldwell, 1996: 32-47).

In the early 1900s, western countries created the first international institutions on species protections, attempting to ensure controlled harvesting of protected species for local economic and social purposes (Barrett, 2003). That is, protection measures that were taken to protect sets of deteriorating animal populations were linked to issues of human survival and development: in current language, actions were taken to preserve a sustainable yield. One of the first such agreements – the Convention between the United States and other Powers providing for the Preservation and Protection of Fur Seals – was negotiated in 1911 to protect fur seals of the North Pacific for the purpose of securing conditions that would allow long-term continuous hunting for economic gains and prosperity (Hayden, 1942). Also other international efforts on ecological protection were initiated in the early 1900s. One of the goals of the InternationalenKonferenzfürNaturschutz [International

Conference on Nature Protection], which was organized on an initiative by the Swiss government, in Bern, in 1913, where delegates from sixteen western countries participated, was to accomplish international regulation for nature protection. These regulatory efforts were directed toward the preservation of threatened species like the musk ox in Greenland and reindeer on Spitzbergen (Frank *et al.*, 2000). In addition, the delegates at the Bern meeting resolved to establish an information agency for world-wide nature protection (Conference for the Establishment of the International Union for the Protection of Nature, 1948; Barthelmess, 1972: 121-28; Conwentz, 1914; Nash, 1992: 358; Linnér and Lohm, 1999: 123-27). These plans, however, were postponed nine months later due to the outbreak of World War I. In the years immediately prior to World War I, European conservation societies developed plans for wildlife protection both in their own countries and their colonies. After World War I, European colonial powers continued to impose species protection measures and created protected areas in their foreign colonies and territories. Minor international agreements were reached on the protection of whales and of African wildlife. Further, the meeting in 1933 of the Convention for the Protection of African Flora and Fauna, which historian Roderick Nash has labeled as the "high point of institutionalized global nature protection before the Second World War," promoted the establishment of national parks and regulations on trade in animal trophies (Nash, 1992: 359). Yet, the Convention failed to acknowledge the close international relationship between human development and species protection in both the North and the South, which become an important topic for international conservation first a decade later. In addition, minor international action was taken shortly before World War II on pollution from shipping (Caldwell, 1996: 235).

After the end of World War II, international conservation efforts were expanded to include broader social aspects of natural resource use and management. This expansion of conservation efforts from selected species protection and the creation of national parks to a more comprehensive focus on natural resource management and conservation, led by a small but growing number of scientists, conservation advocates and policy makers, was prompted by a growing concern about sharp increases in world population and continuing industrialization in an increasing number of countries. Population increases, an accelerated depletion of non-renewable natural resources, and a growing generation of air and water pollution were having important social and economic implications, which resulted in increased attention to the linkage between environment and development issues. Many Western countries, with increasing participation of Eastern and Southern countries, continued to create international institutions for ecological protection and natural resource management. In turn, these institutions promoted increased attention to environment and development issues. For example, in June 1947, the Swiss League for the Protection of Nature sponsored a conference for conservationists, held in Brunnen, Switzerland. Prominent conservationists from around the world met to confer about how future international nature protection should be organized. Delegates at the meeting believed that a new organization should be created, the International Union for the Protection of Nature (IUPN), which should not be concerned solely with the preservation of wildlife, but address conservation issues in the context of accelerating depletion of natural resources and global population increase. IUPN was formally constituted in Fontainebleau, France, in 1948 (Swiss League for the Protection of Nature, 1947: 15-17; Nash, 1992: 361; Conference for the Establishment of the International Union for the Protection of Nature, 1948).

The Preamble of the Charter of the IUPN stated that human civilizations' dependency upon renewable resources demanded the preservation of the entire world's biotic environment, and IUPN officials and delegates claimed that there were social, educational, cultural and economic reasons for protecting nature (International Union for the Conservation of Nature and Natural Resources, 1974: 17-21). In doing so, IUPN sought to take a much broader approach to natural resources management compared to most existing conservation practices (International Technical Conference on the Protection of Nature. Proceedings and Papers, 1950: vii). Founded by conservationists but welcoming government officials, IUPN worked to became an organization that acted as a link between the scientific community and political institutions (Dalton, 1994: 34). In this respect, IUPN was an early example of a "boundary organization" (Guston *et al.*, 2000), seeking to facilitate the transfer of scientific knowledge to policy makers.

The expansion of the international political focus on natural resources, their management and their connections to human development also reflected a change in world politics, as the Cold War began in the late 1940s. By the middle of the 20th century, modern airplanes, ships and missiles forced national security interests to take a global perspective. In addition, the global expansion of the capitalist economy and international trade resulted in a transition in economic production and its related modes of social and political regulations across countries (Linnér, 2003). One example of the expansion of national security concerns related to the growing political interest in environment and development issues was the U.S. proposal to hold the United Nations Scientific Conference on the Conservation and Utilization of Resources (UNSCCUR). UNSCCUR was held in 1949 at Lake Success, USA, and included 550 governmentally appointed delegates from 48 countries, together with representatives from several non-governmental organizations and so-called independent experts. UN Secretary-General Tryggve Lie opened the conference by declaring that "Today the United Nations is embarking on a new phase of its program to build the foundations for permanent peace" (United Nations, 1950-51). As reflected in this statement, UNSCCUR's focus on natural resources was directly linked to political development and security concerns.

UNSCCUR originated in a proposal by the United States to the United Nations Economic and Social Council (ECOSOC) in September 1946. U.S. President Harry S. Truman saw the issue of natural resources and their management as strategically important to U.S. foreign and security policy, building on an earlier idea conceived by U.S. President Franklin D. Roosevelt, (Caldwell, 1996: 51). Through the Point Four Program, named after the fourth point in his 1949 inauguration speech, President Truman wanted to make American science and technology accessible to less economically, technically, and scientifically developed countries by providing aid to these "underdeveloped areas" (Truman, 1964: 11415).

The reason for this proposal strongly reflected Cold War politics and U.S. geopolitical concerns: Truman argued that dire social and economic conditions in developing countries were a hazard not only to themselves and their populations, but also to richer countries, as "underdevelopment" in the least economically, technically, and scientifically developed countries could give rise to social unrest in these countries that could lead to dangerous international political instability in the Cold War system. Truman considered UNSCCUR an important means to promote international stability and U.S. security interests with regard to the developing world (Linnér, 2003: 84-85). Expressing similar beliefs as those that motivated President Truman to formulate his Point Four program, but for different underlying reasons, the first director of the United Nations Food and Agricultural Organization, Lord John Boyd-Orr, who received the Nobel Peace Prize in 1949, stressed that the world by the mid-20th century had become so small that the only hope for human survival was to initiate extensive global cooperation and action to abolish poverty. He argued that countries shared a responsibility to work together to raise the standard of living among people in the least economically and technically developed countries (Boyd-Orr, 1953; Linnér, 2003).

When the United States presented the UNSCCUR conference proposal to ECOSOC, the Cold War was just beginning. One year before the conference, Andrei Gromyko, then Deputy Minister of Foreign Affairs and Soviet representative to the United Nations, announced that the Soviet Union had no objections to make to the preliminary program. Nevertheless, as tensions between the two superpowers escalated, Soviet Union backed out of the conference in the spring of 1949. Czechoslovakia and Yugoslavia were the only Eastern European countries that attended UNSCCUR. In addition, the Guomindang regime in China expressed support for the conference, but did not attend because the then ongoing Chinese civil war (Linnér, 2003: 45). The notion that the world must be regarded as a unity and that the wise use of the earth's resources must be planned through international cooperation was central at UNSCCUR. The distress in the developing world became an important argument for conservationists by which they could appeal for conservation of the world's natural resources. The well-known American conservationist Fairfield Osborn expressed this notion in his introductory address at UNSCCUR. To Osborn, it was an astonishing fact that the newlyfounded United Nations was able to gather scientists from all around the world to confront the rapid deterioration of state of the world's resources. He regarded UNSCCUR as a manifestation of that countries realized "with a new and piercing clarity" that their future prosperity was intimately linked to conditions in countries on the other side of the globe (United Nations, 1950-51).

In a limited way, UNSCCUR became a forum for developing countries to critique industrialized countries for exploiting developing countries as a source for cheap raw material (Linnér, 2003: 48-50). These views, however, largely disappeared in the majority of the reports on new technical means to extract better and more of earth's natural resources. Instead, the more radical conservationists who focused more on global distribution issues were forced to organize an alternative conservation conference, arranged by UNESCO and IUPN. This meeting, the International Technical Conference on the Protection of Nature (ITCPN), included 250 participants from non-governmental organizations in 58 countries, and was held at same time as UNSCCUR. The conference aimed to develop a worldwide program for the conservation of food resources.

At ITCPN, the General-Secretary of IUPN, Jean-Paul Harroy, promoted a new conceptualization of natural protection. Harroy argued that the time had passed when conservation should be directed merely towards constructing regulations and setting apart nature reserves to safeguard biota for mainly aesthetic reasons. The precarious situation of a rapidly growing world population demanded that conservation should be directed towards the political management of natural resources, according to Harroy (International Technical Conference on the Protection of Nature, 1950). Many of the issues that were discussed at ITCPN continued to be subject to discussion and research after the meeting, although mounting tensions between East and West slowed down multilateral cooperation on natural resources management. Nevertheless, countries concluded the Convention on Fishing and Conservation of the Living Resources of the High Seas in 1958, seeking to avoid over-fishing.

After World War II, there was a growing interest among the leading industrialized countries in the use of science and technology relating to natural resources management and the study of the global biosphere (Clark *et al.*, 2001). Several national research institutions and governmental bodies, inter-governmental organizations, and non-governmental organizations participated in these debates and efforts, collected data, and compiled reports (Strong, 1972). This intensification of scientific research and technical development took place within the context of EastWest competition. As discussed above in connection with Truman's Point Four Program, there were highlevel political attempts in the United States and elsewhere to promote and disseminate new science and technology for improved natural resource management to support economic development and address geopolitical security concerns. In part, American and Soviet governments hoped that the new scientific research and technology would give them practical advantages in their competition with each other (Weart, 2003).

Increasing research activities in the 1950s stimulated the creation of large international collaborative research programs on issues relating to the natural environment. This was an area in which there was often east-west cooperation. Following a suggestion by the U.S. scientist Lloyd Berkner, the International Council of Scientific Unions (ICSU) in 1952 proposed a series of global geophysical activities for the period July 1957-December 1958 under the so-called International Geophysical Year (IGY). The IGY was modeled on earlier International Polar Years (1882-83 and 1932-33), and was intended to allow scientists from around the world to take part in a series of coordinated observations of various geophysical phenomena.

The IGY included representatives from 67 countries. IGY activities spanned the globe, using the latest scientific tools. Although much work was carried out in the Arctic and equatorial regions, special attention was given to the Antarctic. IGY Antarctic research on ice depths yielded radically new estimates of the Earth's total ice content, and also contributed to improved meteorological prediction, advances in the theoretical analysis of glaciers, and better understanding of seismological phenomena in the southern hemisphere (Fleagle, 1992). In addition, scientists such as Charles Keeling acted under the IGY to forward the science of climate change and monitoring of carbon dioxide (Weart, 2003). Another international research program, the International Biological Program (IBP), ran between 1959 and 1974 and involved experts from 58 countries. The IBP included research projects in areas such as the ecology of natural resources, population problems, and nature conservation. One of the IBP's main motives (and also a driving factor behind the launching of the IGY) was to take advantage of technology emerging out of the space race such as remote sensing to look at the well-being of the planet. This new technology gave an opportunity to study global natural processes for the first time (Worthington, 1975; Blair, 1979; Fries, 1984; McIntosh, 1985; Kwa, 1987; Worster, 1994). In 1966, the first photos of Spaceship Earth were taken by the U.S. Lunar Orbiter from the vicinity of the moon (Jasanoff, 2004).

In September 1968, the Intergovernmental Conference of Experts on a Scientific Basis for a Rational Use and Conservation of the Resources of the Biosphere (the so-called Biosphere conference) was held under the auspices of UNESCO (Caldwell, 1996: 53-54). Caldwell has noted that the "twenty years between the resources (1949) and the biosphere (1968) conferences mark as fundamental a change in perceptions of international responsibilities for the global environment as any that have occurred since the establishment of permanent international organizations" (Caldwell, 1996: 53-55). Unlike at UNSCCUR in 1949, recommendations from the Biosphere conference called for comprehensive international and domestic environmental management and stressed that environmental issues were closely connected to societal issues.

The IBP and the Biosphere conference led in part to the Man and the Biosphere (MAB) program, created in 1971 under UNESCO in collaboration with the United Nations Food and Agriculture Organization (FAO), the World Health Organization (WHO), the World Meteorological Organization (WMO) and ICSU. The MAB program is still active as an independent research programme. The programme stimulates interdisciplinary research and training in natural resource management, and promotes the involvement of science and scientists in policy development on issues relating to biological diversity. In these capacities, the MAB programme seeks to promote sustainable development through conservation and wise use of biodiversity.

The 1963 UN Conference on the Application of Science and Technology for the Benefit of the LessDeveloped Areas was the first UN attempt to discuss the role of science and technology for development (rather than natural resource management). The conference proposed the transfer of technology from industrialized countries to developing countries so that developing countries could accelerate the transformation of their economies and their social and economic development. The conference also resulted in the creation of several UN bodies for science and technology advice, such as the Committee on Science and Technology for Development under ECOSOC, the Advisory Committee on the Application of Science and Technology for Development, and the Office of Science and Technology within the UN Secretariat (United Nations, 2003: 7-8). Two issues increased dramatically in international political importance after World War II and served to put environment and development concerns on the political agenda: population growth and pollution issues. This importance can be seen in the fact that they quickly became central in the emerging environment and development discourse. Many regarded issues of rapid population growth (especially in developing countries) and increasing air and water pollution (mainly in industrialized countries) to warrant the highest level of political attention, both domestically and internationally.

Global population growth became increasingly linked to natural resource management issues, environment issues and development issues after World War II. Roughly outlined, rapid population growth, predominantly in developing countries, was feared to have adverse effects not only on the environment, but also on the possibilities of development. As developing countries' populations grew, environmental strains would increase, as well as the demand for natural resources – limited natural resources which were needed for economic development in both developing and industrialized countries. To address these issues, the United Nations held major population conferences in Rome in 1954 and in 1964, discussing connections between natural resources, population growth and development, as well as how these issues such be addressed by the international community. Presenting such arguments, Paul Ehrlich's The Population Bomb (1968), which examined risks associated with a quickly accelerating population, gained widespread public and political attention.

Another highly publicized study was produced by the so-called Club of Rome, which attempted to examine the limits to growth in a world with finite resources (Meadows *et al.*, 1972). To that end, they used computer models to predict future conditions and trends based on past and present data. The study looked at five major trends of global concern: (i) accelerating industrialization; (ii) rapid population growth; (iii) widespread malnutrition; (iv) depletion of non-renewable resources; and (v) deteriorating environment. These trends were treated as dynamically interacting and exhibiting exponential growth. Warning about a faith in technology as the single solution, the authors presented three major conclusions from their computer modeling exercise: First if the current growth trend in the five areas continued unchanged, the limits to growth on planet earth would be reached within the next 100 years. Second, it is possible to alter current trends and establish a condition of ecological and economic stability that is sustainable. Third, the sooner efforts begin to reach a condition of ecological and economic stability the greater the chances of success.

Emerging pollution problems, predominantly in industrialized countries, that drew both public and political attention in the 1960s included cases of mercury poisoning in Minamata, Japan (giving name to the so-called Minamata disease); oil pollution around Santa Barbara in California; smog problems in New York; and pollution of the river Rhine. Rachel Carson's best-selling Silent Spring (1962) started much discussion on the growing use of pesticides. International concern about industrial chemicals increased significantly in 1968, when many people in Yusho, Japan, were poisoned after eating rice contaminated with high levels of polychlorinated biphenyls (PCBs) that had leaked from a heat exchanger. The Organization for Economic Cooperation and Development (OECD) was the first international organization to pay serious attention to the environmental and human health impacts of hazardous chemicals, seeking to improve and harmonize domestic technical, scientific, and policy measures among its member states (*i.e.* industrialized countries) (Downie, Krueger and Selin, 2004). Marine pollution was another pollution problem that gained multilateral attention and prompted the formulation of pollution prevention treaties. Political cooperation on pollution led to the 1954 International Convention for the Prevention of Pollution of the Sea by Oil. In addition, the First United Nations Conference on the Law of the Sea (UNCLOS) was held in Geneva in 1958. In 1961, the United Nations Economic Commission for Europe, which contained members from both the East and the West, held its first meeting on water pollution. Both population issues and pollution issues were part of the international environment and development discourse that was developing in the 1960s, partially as a result of scientific and technological advances in the study of ecology and environment. As efforts were made in the United Nations and elsewhere to formulate multilateral responses to the challenges of rapid population growth and pollution, these issues became part of East-West politics and also signaled the emergence of a more marked North-South division. The political dimension of these issues became clear as countries began preparing for the first global meeting on environment and development, the 1972 UN Conference on the Human Environment.

18.4. The Impact of Foreign Private Investment (FPI) on Capital Formation in Nigeria

In the late 1970s and early 1980s, most developing countries of Africa (including Nigeria) experienced unprecedented and severe economic crisis. This crisis manifested itself in several ways such as persistent macroeconomic imbalances, widening saving-investment gap, high rates of domestic inflation, chronic balance of payment problems and huge budget deficit(Akpokodje, 1998). Although different reasons have been adduced for the slowdown of these economies, Greene and Villannueva (1991) attribute the problem to the decline in investment rates in the affected economies. In Nigeria, for example, Akpokodje (1998), maintained that domestic investment as a ratio of gross domestic product (GDP) declined from an average of 24.4% during the 1973-1981 period to 13.57% during 1982-1996 period. The average investment rate during the 1982-1996 period implied that the country barely replaced its dwindling capital. In the same vein, private investment rate depreciated from 8.6% in 1973-1981 period to 4.2% in the 1982- 1996 era. Due to the fact that investment determines the rate of accumulation of physical capital (otherwise called capital formation), it then becomes a vital factor in the growth of productive capacity of the nation and contributes to growth generally. It is in the light of this that prominence is being attached to increasing the magnitude of real asset investment in the economy.

In particular, central to the less than satisfactory growth registered by countries of sub Saharan Africa is low level of investment as a result of low domestic savings. Attracting foreign investment is therefore crucial from a number of standpoints and of course, there is never shortage of theoretical arguments (Chete, 1998). First, consistent and regulated inflow of foreign investment provides an important source of foreign exchange earnings needed to supplement domestic savings and raise investment levels. Second, import substituting investment would serve to reduce the import bills as investment in export industries will directly increase the country's foreign exchange earnings.Some other benefits might also accrue from increased foreign investment. These include the creation or rather expansion of local industries to supply inputs to the newly established plants; a rise in the overall level of domestic demand to boost incomes and, through taxation, state revenues; and the transference of labour (human capital) skills and technology. Yet another set of benefits arises from the forecasting of efficiency in the domestic economy, an effect that might even occur prior to the anticipated investment flows (Chete, 1998).

Most probably due to these overwhelmingly attractive theoretical arguments in support of foreign investment, government authorities in Nigeria have often articulated a plethora of incentives aimed at attracting foreign investment. For example, the *New Industrial Policy* published in 1988 embodies some Foreign Direct Investment (FDI) provisions which represent a dramatic departure from the past policy (see Chete, 1998, for details)Besides, the need for external capital inflow arises, when desired investment exceeds actual savings. They are necessary also owing to investments with long gestation periods that generate non-monetary returns; growing government expenditures that are not tax-financed and when actual savings are lower than potential savings owing to repressed financial markets, and even capital market flight (Ogamba, 2003).

Several variables which create dependence on foreign capital have been identified in the literature. They could be classified into fluctuating variables such as exports, imports and invisible; offsetting variables like debt service and reserve creation, and rigid variables which include minimum level of imports, stage of economic development and exportable surplus (Ogamba, 2003). External capital flows could also be non-debts creating flows (as in official transfers or grant in aids and direct investment flows), debt creating flows (as in official development finance), commercial bank loans and international bond offerings; or could equally be a hybrid, for example, foreign portfolio investments and international equity offerings. Of late, Nigeria has embarked upon several trade liberalisation policies so as to free FDI flows into the country (Adegbite and Owuallah, 2007).

The literature is replete with evidence that private investment in most developing countries is more directly related to growth than public investment (see Akpokadje, 1991; Serven and Salimano, 1991; Khan and Reinhart, 1990). Accordingly, it is now widely accepted that the expansion of private investment should be the added impetus for economic growth in developing economies (Chhibber and Dailami, 1990). Many developing countries have over the years relied very much on the inflow of financial resources from outside in various forms, official and private capital flows as well as direct foreign investment, as a means of speeding up their economic development (Olaniyi; 1988, Odozi, 1995; Ekpo, 1997 and Uremadu, 2006). However, these countries have shown preference for direct foreign investment because they regard direct foreign investment as a means of counteracting the sluggish trend in official and private portfolio capital flows.

Generally, capital from outside can be very helpful in speeding up the pace of economic development and can act as a catalytic agent in making it possible to harness domestic resources particularly in a developing country. But foreign capital, no matter how large the inflow, cannot absolve a recipient country from the task of mobilising domestic resources. Foreign inflows can, at best, be complementary to domestic savings. In developing economies experience has shown that foreign capital alone cannot create any permanent basis for a higher standard of living in the future. Rather it complements domestic savings. Therefore greater dependence on internal sources of finance facilitates the successful implementation of any planned economic development in a country (Agu, 1988 and Uremadu, 2006). The same low and fluctuating trend has been exhibited by the gross domestic investment also known as Gross Fixed Capital Formation (GFCF). Certainly, macroeconomic policies: monetary, fiscal and exchange rate, have a bearing on the investment behaviour in a country (Likewelile, 1997, Ghuma and Hadjmichael, 1996; World Bank, 1994 and Akpokodje, 1998), but the impact of these policies on private investment behaviour in Nigeria is still largely unclear. This book therefore explores the association existing between capital formation and other macroeconomic indicators of interest in the pursuant of macroeconomic policies in Nigeria.Specifically, it seeks to determine the impact of cumulative foreign private investment on capital formation and growth in Nigeria. It will also highlight the complementary role played by it to gross domestic savings towards filling the existing savings - investment gap in a bid to achieve desired investment goals and/or growth objectives in Nigeria in the years ahead.

The preponderance of empirical studies that have explored quantitatively the determinants of foreign direct investment have concentrated more on economic than other factors. In particular, each of the authors, in his regression equations included those determinants the or she considered personally appealing. In what follows, we survey some of these empirical investigations. A leading proponent of the economic approach to the determinants of foreign direct investment is found in (Dunning, 1973). On the strength of studies by scholars based on international production, he identifies three sets of influences on foreign direct investment to include the following:

- ☆ market factors such as the size and growth of the market measured by the gross national product (GNP) of the recipient country;
- ☆ cost factors such as the availability of labour, low labour costs and inflation; and
- ☆ the investment climate as measured by the degree of foreign indebtedness and the state of the balance of payments (Chete, 1998).

In another study, Dunning (1981), develops an eclectic theory of international direct investment based on the theories of industrial organization of location of a firm. Nonetheless, the treatise of this later study does not directly concern the subject in hand. Agarwal (1980), clarifies the determinants of foreign direct investments using two political factors of political stability and the threat of nationalization, in conjunction with a variety of economic factors such as investment incentives, the size and growth of the recipient's market, its degree of economic development proxied by infrastructure, market distance and economic stability in terms of inflation, growth and balance of payments. In his extensive survey of the literature on the determinants of foreign direct investment, he finds mixed evidence with respect to the impact of political instability.

Lewis (1978), lays emphasis, to some extent, on political factors too. He tested the dual hypotheses that economic considerations are the prime determinants of foreign investment flows and that political variables are of secondary importance. His model uses a step-by-step regression for 25 developing countries from three continents: Africa, Asia and Latin America to establish that economic variables are more important than the political ones. All the above studies except, Dunning (1981), were pre-occupied with the determinants of foreign direct investment in developing countries. A respectable number of studies have also been conducted for developed countries particularly for the United States and the European Community (*e.g.*, see, Scapelanda and Balough, 1983 and Lunn, 1980). The authors established similar findings.

Back home in Nigeria, significant scholarly effort has gone into the study of the role of foreign direct investment in the Nigerian economy. For instance, Oyaide (1979), provides an excellent documentation of works conducted under the umbrella of Nigerian Economic Society (NES). What follows draw substantially from this brilliant summary as reported in (Chete, 1998). The preoccupation of Edozien (1968), is on the linkages generated by foreign investment and their impact on Nigeria's economic development. Specifically, he contends that foreign investment induces the inflow of capital, technical know-how and managerial capacity which interactively will accelerate the pace of economic development, while attenuating the pains and uncertainties that come with it. Furthermore, he observes that foreign direct investment could be counter-productive if the linkages they spur are neither needed nor affordable by the host country.

Conclusively, he suggests that a good test of the impact of such investment on Nigeria's development is how rapidly and effectively it fosters local enterprises to innovation. In a related study, Largely (1968), posits that foreign direct investment has both benefits and repercussions in the context of Nigeria's economic development. While FDI could accelerate growth through the infusion of new techniques and managerial efficiency, she however warns that it could worsen the balance of payments position. She stopped short though, of elaborating the channels through which this can be actualised. Foremost, Olakampo (1962), has alluded to this negative fall-out of FDI when he argues that foreign aid in the form of direct investment and portfolio investment generally imposes a burden of repayment in form of capital outflows on the recipient country. Oyaide (1977), concludes, using indices of dependence and development as mirror of Nigeria's economic performance, that direct foreign private investment (DFPI) engineers both economic dependence and economic development. In his view, DFPI continuously causes and catalyses a level of development that would have been impossible without such investment albeit, at the cost of economic dependence. Olopoenia (1983), explores the role of foreign capital inflow in the development processes of underdeveloped countries via its impact on savings. He fails, however, to reach unambiguous conclusion, contending that the effect of foreign investment on saving depends on the savings hypothesis used. Additionally, Oseghae and Amenkhienan (1987), examines the relationship between oil exports, foreign borrowing and direct foreign investment in Nigeria, on the one hand, and the impact of these on sectoral performance, on the other hand. They surmise that foreign borrowing and FDI impacted negatively on overall GDP but positively on three main sectors (manufacturing, transport and communication, and finance and insurance).

Elsewhere, Olaniyi (1988) investigates the impact of direct foreign capital on domestic investment to ascertain its overall contribution to the enhancement of the domestic savings capacity in Nigeria. His model of domestic savings and investment financing in Nigeria empirically tested the impact of FDI on the level of domestic savings and investment. His results conform that domestic savings is by far more relevant in determining investment growth than foreign capital inflows in Nigeria. At best, the latter complements the former. This view has also been confirmed by the works of both Uremadu, (2006) and Adegbite and Owuallah, (2007). Evidently, from the Nigerian studies reviewed so far, there seems to be a dearth of research explicitly devoted to the impact of foreign private investment neither on capital formation nor on the determinants of foreign private investment in the Nigerian scene proper thereby re-enforcing the need for this current effort.

With respect to the foreign studies surveyed earlier, it is apparent that, in general, empirical exploration of the determinants of foreign direct investment has not been definitive on what are the main factors. In particular, it is largely inexplicit what role the economic factors play relative to political factors. Curiously, the less than satisfactory state of research on this theme can be traced to issues of content and of statistical methodology. In relation to content, the studies show wide variance of economic and political factors used as arguments. And no convincing reason rooted in the theory is advanced for the inclusion or exclusion of particular variables (Chete, 1998). Thus, a drawback in most of these works is that their empirical estimation and the variables used as causal factors are not guided by theoretical considerations but mere *ad-hockey* as postulated in (Chete, 1998). The conclusion therefore is trite that the existing state of research shows serious conceptual and statistical weaknesses, providing further impetus for this study.

Keynes (1936), pioneered the discovery of an independent investment in the economy, in contrast to, the widespread belief (that is, the Wicksellian loan market) that all available saving is automatically invested so far as the interest rate is attractive. Keynes' main contention was that investment is a function of the prospective marginal efficiency of capital relative to some interest rate which reflects the shadow cost of the invested funds. According to Keynes (1936), because of incomplete and uncertain information about private investment volatility in the future, potential investors would depend on their "animal spirits" in making their investment decisions rather than a rational calculation of an inherently intermediate distant future (Chete, 1998).

Investment theories that followed the tradition of the Harod Domar growth models emerged in the 1950s and 1960s. This was the precursor to the familiar accelerator theory. This theory posits investment as a linear function of changes in output derived from a fixed proportion of production technology. Thus, given an incremental capital-output ratio, it is easy to compute the investment requirements needed to achieve a given output growth target. In his model, profitability expectations and cost of capital considerations are ignored in the determination of investment.

The Neo Classical Approach to investment was next in line. Mainly spurred by the desire to obviate the shortcomings of the Harod Domar formulation, particularly in its simplistic assumptions, this approach introduces factor substitution in the derivation of the demand for capital from the firm's cost minimization problems. Consequently, the desired capital stock is shown to depend on the rental cost of capital (which, in turn, depends on the price of capital goods, the real interest rate and the depreciation rate) and the level of output. This approach too has been attacked on account of inconsistency of the assumptions of perfect competition and exogenous output, the inappropriateness of static expectations and the introduction of delivery lags in an *ad hoc* manner.

Tobin's "Q" theory of investment of 1969 is an alternate formulation of the investment function. The theory postulates that the ratio of the market value of the existing stock of capital to its replacement cost (otherwise termed Q ratio) is the force driving investment. Tobin, subsequently elaborated two reasons why Q may differ from unity which include delivery lags and increasing marginal costs of investment.

Abel (1981) and Hayashi (1982), both, in separate studies, attempt a reconciliation of the Neo-Classical and Q approaches to investment. By showing that the latter follows from the firm's optimal capital accumulation problems under adjustment costs, they proved that what drives investment is marginal Q, that is, the ratio between the increase in the value of the firm due to the installation of an additional unit of capital and its replacement costs. However, marginal Q may not be observable as it will generally diverge from the observed average Q (which essentially is the market value of existing capital in terms of new capital), except under competitive equilibrium and constant returns to scale. Both will also diverge if firms confront quantity constraints in real asset or financial markets. In such a situation, average Q will, at best, provide some and not all of the information required to make investment decisions.

This disequilibrium approach to investment is associated with Malinvaud (1980, 1982) and Sneessens (1987). In this respect, investment is a function of both profitability and output demand considerations. Malinvaud (1982), splits investment decision into two stages: the decision to expand the level of productive capacity, and the decision about the capital intensity of that additional capacity. The latter depends on profitability variables like the relative cost of capital and labour. By contrast, the capacity decision relates to the degree of capacity utilization in the economy as an indicator of demand conditions. In Sneessens (1987), net investment is a positive function of the gap between actual and long-run equilibrium capacities. This, in turn, is a reflection of differences between actual an equilibrium rates of capacity utilisation and between actual and equilibrium mark-up rates. Therefore, investment depends on both profitability (discrepancies between actual and equilibrium mark-up rates) and on sales constraints (discrepancies in rates of capacity utilisation). By implication, investment decisions may be reached in a setting in which some firms confront current and expected future sales constraints, a crucial departure from both the Neo-Classical (Jorgenson) and the Q (Tobin) models.

There exists an expanding literature on the effects of financial constraints on investment, for example, as seen in (Fazzari, Hubbard and Peterson, 1988a, 1988b; Caloris and Hubbard, 1989; Mayer, 1989 and Mackie and Mason, 1989). The central argument here is that internal finance (retained earnings) and external finance (bonds, equity or bank credit) are not perfect substitutes. The discrepancy in the cost of different sources of financing is due to asymmetric information where bankers in the capital market cannot evaluate accurately the quality of firm's investment opportunities; thereby leading to existence of opportunity cost of internal finance generated from cash flows and retained earnings. According to this view, investment will be very sensitive to financial factors such as the availability of internal finance or the access to capital markets (Chete, 1998).

Finally, another feature of investment in developing countries of sub-Saharan Africa (SSA) is the high import content of capital goods. This buttresses the contention in the two gaps model (Chenery and Bruno, 1962 and Bacha, 1982), that the lack of foreign exchange may constitute a major constraint to sustain high rates of investment and growth in developing economies. Therefore in countries like Nigeria where domestic and foreign capital goods are highly complementary, the lack of foreign resource to import machinery and equipment will always constitute an impediment to growth. In other word, foreign exchange is a crucial determinant factor in capital formation among developing countries of Africa.

19
Income Distribution and Development

19.1. Income Distribution and Development

The distribution of income within a society is of enormous importance. It influences the cohesion of the society and, for any given level of GDP, determines its poverty level. Some relatively high-income economies have very unequal income distribution with the result that there are large cleavages in society and high levels of poverty, as in Brazil. Other countries with more equal income distribution have less poverty and there is a sense of fairness within the society which makes for political stability, as in Costa Rica. The sensitivity of poverty to growth depends on a country's income distribution; for example, a 1 per cent growth rate of GDP leads to a 0.21 per cent reduction in poverty in Zambia, if distribution is unchanged, compared with a 3.4 per cent reduction in Malaysia (Sen, 1995). There is also considerable evidence that the distribution of income has a significant influence on the rate of growth, with more equal societies growing faster than less equal ones. Moreover, the average health status of a society depends on its income distribution, to that countries with more unequal distributions experience lower life expectancy. An equitable distribution of income, as well as the achievement of social goals, are, therefore, essential aspects of development, over and above economic growth.

This book aims to explore the connections between income distribution and economic growth, and to identify some policy conclusions emerging from the analysis. There have been many investigations of the relationship between income distribution and development, starting with a classic paper by Kuznets, who argued that income distribution was generally relatively more equal at low levels of income in the early stages of development, became more unequal as development proceeded, and finally a reverse move took place so that income distribution became more equal again as countries approached the levels of incomes of the developed countries. The work of Kuznets, and others, identified correlations between levels and growth of per capita income and income distribution. Behind these correlations lie two possible types of causality: first, how growth affects the distribution of income; and secondly, how distribution affects growth. Both will be investigated, before exploring recent trends in income distribution.

19.1.1. Some Important Definition Issues

Key issues concern distribution of what, among whom and within which unit.

(*a*) *Distribution of what?* In almost all discourse, the focus is on the distribution of current monetary (or private) income – normally pre-tax and subsidy but sometimes post-tax and subsidy. Even within the 'income' paradigm of welfare, one needs to extend this to include future income (*e.g.* by adding current asset distribution). But alternative approaches to well-being suggest the need for concern with one, other, or all of the following: the distribution of social income (*i.e.* goods and services provided by the State) as well as private income; distribution of capabilities or functioning; of basic needs goods and services; or of human development achievements. A broader approach to distribution is needed, whether it is a matter of assessing distribution in a society from the perspective of well-being, or from that of investment (*i.e.* the impact of current distribution on growth). For example, inequality in access to education, which is acute in many societies, is a major influence on future household income, and may also affect the rate of growth, since in many occupations those deprived of education are unlikely to be as productive as the educated. Inequality in access to health services can be much more important than inequality in private incomes, since life itself may depend upon it. Nonetheless, most of the literature focuses on the distribution of private incomes.

(*b*) *Distribution among whom?* The appropriate level of analysis depends on why the information is needed. If the aim is to assess well-being, and an individualistic approach to well-being is adopted – as with utilitarian and capability approaches – then the individual is the appropriate level. But much of the data is collected at a household level, as it is difficult to get information at the level of the individual. Frequently, household and individual information is used interchangeably, but there has been some effort to correct household data, in order to translate it into information about individuals, for example, by allowing for intra-household distribution (and sometimes correcting income to allow for the needs of people of different ages or genders within the household). *Vertical* measures of distribution among households or individuals. Horizontal distribution is one of the major causes of conflict between groups. It can also be directly relevant to individual well-being, where people identify strongly with the group to which they perceive themselves as belonging.

For other purposes, such as for North-South negotiations, the appropriate measure of inequality may be average differences among nations.

(*c*) *Distribution within which unit?* Another issue is the *unit* within which inequality is assessed. Conventionally, this unit is the *nation*, the obvious rationale being that this is the major policy-making unit. But some policies are made at the local level, so that the local administrative unit would be the relevant one. For purposes of aid and other international policies, the global level is appropriate, and for intra-regional policies, distribution within the region. In practice, almost all the literature relates to *private income distribution among individuals within a nation*, thereby excluding many important issues. Much of this paper will do likewise, but will return to some of the broader definitions at the end.

19.1.2. How Growth Affects Distribution

As noted earlier, in 1955 Kuznets famously propounded the view that there is an inverted "U" curve relating levels of per capita income to income distribution, with income distribution first becoming more unequal, and at a later state more equal, as per capita incomes rise. Kuznets derived this from cross-country evidence. Historical work on the changing income distribution in industrialized countries also provided supporting evidence (Paukert, 1973). But it should be noted that the Kuznets work relates to *levels* of income per capita, not to the *growth* rate. Moreover, further work on the Kuznets curve has found the relationship weak, as it is dependent on the precise functional form adopted (*e.g.* Anand and Kanbur, 1993a; Deininger and Squire, 1998). Bourguignon concludes: "If there is any parabolic relationship between income inequality and GDP per capita across countries. it is probably very weak and unstable over time. [and] longitudinal data. seem to suggest that there is much freedom in the way distribution in a given country may change over time" (Bourguignon, 1995: 47). Deininger and Squire (1998) also find 'virtually no support' for the Kuznets hypothesis. However, there is no uniform agreement on this. Several investigations have found some support for the Kuznets hypothesis (*e.g.* Oswang, 1994; Ali, 1998; Milanovic, 1994; as well as Fishlow, 1995). Despite its fragile empirical foundations – made even more so by the recent increase in inequality in developed countries – the Kuznets curve has been widely accepted, and sometimes used as an excuse, for taking no action on income distribution, on the assumption that the natural laws represented by Kuznets will unavoidably be realized. It may, of course, be that there are 'natural laws' leading to a Kuznets relationship in a laissez-faire development process, but these can be countered by policy which explains the many exceptions to the curve. It is in this spirit that it is worth briefly surveying the explanations that have been put forward for the Kuznets relationship:

(i) In the case of a dualistic economy, with a low productivity, more egalitarian agrarian sector, and a high productivity, less egalitarian industrial sector, development causes a sectoral shift to occur. Consequently, inequality rises, both because of the differences in average incomes between the sectors and because people are shifting from a less to a more egalitarian sector. But when the whole economy becomes part of the modern industrial sector, the inequality arising from the between-sector differences disappears, and when full employment is reached, income inequality within the modern sector also diminishes. This explanation, termed the Kuznets process, was developed into a formal model by Anand and Kanbur (1993b);

(ii) Another explanation, derived from the Lewis growth model, is that productivity and incomes growth is confined to the modern sector, where the profit share rises, while in the stagnant traditional sector incomes remain low (and may even fall as population growth occurs), so that between-sector inequality rises;

(iii) A third explanation attributes the process to an initially unequal distribution of assets, which contributes to rising inequality, as those with more assets also accumulate more; but, eventually, the rate of return to capital falls and the unequalizing effect of capital is offset by an equalizing effect arising from labour incomes.

Empirical work on growth (as against levels) of per capita income, however, shows *no* relationship between growth rates and inequality (Ahluwalia, 1976); and recent work confirms this (Bruno, Ravallion and Squire, 1995; UNCTAD, 1997). Histories of individual countries show that in some countries income distribution has worsened over time (*e.g.* Brazil) and in others it has improved (*e.g.* Indonesia in the 1970s). In fact we can observe countries in each of the four possible quadrants representing combinations of growth and changes in income distribution, as shown in Table 19.1.

Table 19.1: Growth and Distribution in different Economies

	High Growth	*Low Growth*
Distribution worsening	Brazil (1960s to early 1990s); Pakistan (1970–1985); China (1980s); Thailand (1970s and 1980s); Botswana, 1970s	Post-Soviet Russia; most eastern European countries, 1980s; Mexico 1980s; Kenya 1980s; Ethiopia 1980s; Guatemala 1970s and 1980s
Distribution improving	Indonesia (1973–1993); Malaysia (1970–1990); Taiwan Province of China (1950–1980); Republic of Korea (1950–1980); Mauritius (1980s and 1990s)	Sri Lanka (1960–1970); Cuba; Colombia, 1980s; Morocco (1970–1984); Trinidad and Tobago (1970s and 1980s)

Source: World Bank report 2014.

The conclusion, then, is that growth is 'distribution neutral', *i.e.* it does not necessarily lead to either a worsening or an improvement in income distribution, and may be consistent with either. Structural factors and policy stances determine countries' experiences.

In the 1950s it was assumed that more unequal income distribution led to higher growth, via higher savings – and possibly incentive effects (*e.g.* Galenson and Leibenstein, 1955). Higher savings propensities associated with more unequal income distribution were variously attributed to the effect of a rising profit share (more, or only, savings out of profits, as assumed by Marx, Kaldor and Lewis), or of more unequal household income distribution (with a Keynesian consumption function). The early choice-of-technique literature (Dobb, 1956–57; Sen, 1968) argued that more capital-intensive techniques should be chosen to maximize surplus and reinvestible funds. From this, the view emerged that countries should grow first and redistribute later. This view was challenged, for example, by Adelman and Morris (1973), who argued that more equal initial income distribution would lead to higher growth.

Recent literature has supported Adelman and Morris; empirical work shows that countries with more equal income distribution have higher growth (Alesina and Perotti, 1993; Persson and Tabellini, 1994; and many others). Owing to data constraints, most work involves cross-country regressions.

There are especially severe data problems in the area of income distribution, as under-declaration of income is common for obvious reasons.5 Here too the robustness of the findings has been questioned, notably by Fishlow (1995), who finds no statistically significant evidence of a relationship between growth and equality, when a dummy variable is introduced for Latin America. However, investigations over time in developed countries have confirmed the relationship between greater equality and higher growth (*e.g.* Panizza, 1999, who investigated growth performance in states in the United States from 1920). However, the very large number of studies finding some relationship gives some confidence in the existence of a positive relationship between equality and economic growth – and certainly refutes the prior and opposite conclusion.

A variety of mechanisms to explain the positive relationship between income distribution and economic growth have been suggested.

(a) One type of mechanism relates to the political economy of more or less equal societies, and how, as a result of political developments, high inequality translates into growth-impeding factors. For example:

(i) It is argued that higher inequality leads to more political instability, more uncertainty, less investment and lower growth (Alesina and Perotti, 1994a; Bertola, 1993; Perotti, 1993; Persson and Tabellini, 1994);

(ii) It is suggested that higher inequality leads to populist redistributive tax policies, more disincentive effects and lower growth (Alesina and Rodrik, 1994; Persson and Tabellini, 1994);

(iii) Higher inequality gives a disproportionate influence to rich groups which lobby for preferential tax treatment, leading to over-investment in certain areas and reducing growth (Bruno, Ravallion and Squire, 1995).

(iv) Including Alesina and Rodrik, 1994; Birdsall, Ross and Sabot, 1995; Bourguignon, 1995; Deininger and Squire, 1996; Sarel, 1997; Larrain and Vergara, 1997.

(v) Deininger and Squire (1996) produced a 'cleaned' data set, excluding 1,200 out of 2,000 observations. But the cleaning process inevitably introduces its own biases; for example, more equal countries are likely to produce more reliable data on income distribution as particular groups have less reason to conceal their incomes. Other explanations relate to the economic effects of higher equality/inequality:

(a) Higher equality of land ownership leads to more labour input, and higher land productivity (*e.g.* Lipton, 1993). There is abundant evidence that more equal land distribution is associated with higher agricultural productivity as well as more equally distributed rural incomes, and in rural economies this accounts for a significant proportion of total incomes.

(b) Higher equality reduces poverty, and leads to more human development (nutrition, education and health), with a more productive workforce, more innovation, etc. (Birdsall, Ross and Sabot, Stewart and Raffirez, 1995; Ranis, 2000).

(c) Higher equality in asset distribution leads to a more even access to credit and information, and more opportunities for the poor to make productive investments (Galor and Zeria, 1993; Deininger and Squire, 1998).

(d) Higher equality leads to larger domestic markets, greater exploitation of economies of scale and hence more industrialization and growth (Murphy, Shleifer and Vishny, 1989).

(e) Higher inequality, on the other hand, leads to higher fertility, since those who are poor and less educated have larger families, and this in turn reduces growth (Benabou, 1996; Khoo and Dennis, 1999; Bloom *et al.*, 1998).

These are all hypotheses, none of which have been thoroughly tested. The political economy explanations mostly rest on an assumption about the behaviour of the 'median voter', which is not relevant in non-democratic societies. Moreover, the statistical evidence supporting the positive relationship between more equality and higher growth appears to apply to non-democratic countries, not democratic ones, according to Deininger and Squire (1998). In addition, the model assumes that more inequality leads to higher government spending, when in fact the reverse appears to be true (Benabou, 1996). The 'human development' explanation (*i.e.* that more equality leads to more education, better nutrition and health, and hence more productive people (see (v) above) is much better supported. There is considerable evidence that a more equal income distribution leads to a greater spread and level of education, as well as improved health and nutrition, and that this in turn brings about higher growth (Birdsall and Sabot, 1994; Ranis *et al.*, 2000). However, this does not seem to be the whole story, as empirical work has found that greater equality has an independent positive impact on growth, in addition to the impact via education levels (Birdsall and Sabot, 1994; Bourguignon, 1995).

Whatever the mechanisms – still subject to investigation – there is broad agreement on the empirical evidence, which shows that more equality is associated with higher economic growth. Moreover,the order of magnitude of the effect is quite high. For example, Bourguignon estimates that a change of one standard deviation in inequality is responsible for half a percentage point of additional growth. Taking the findings of sections II and III together would seem to point to the following fairly optimistic conclusion; namely, that more egalitarian income distribution is better for growth; and that cross-country evidence shows that growth neither increases nor decreases inequality in any systematic way. Since more equal income distribution is desirable as an intrinsic part of the development agenda, as a mechanism for reducing poverty and enhancing human development, *and* as instrumental to growth, the agenda should be to identify which *types* of growth are more likely to improve income distribution and which policies would help bring about egalitarian patterns of growth.

19.2. Egalitarian Patterns of Growth

The distribution of income is the outcome of complex economic processes. Individuals' incomes depend on their incomes from assets and from their own current activities. Income from assets is a function of asset ownership and the rate of return on assets; and income from current activities depends, similarly, on quantity of, and returns to, employment (or self-employment), the returns to employment normally being a function of the individual's level of education and skills. For an economy as a whole, income distribution then depends on asset distribution, distribution of human capital, and returns on each. It is thus not surprising that we observe a large range of income distribution across countries – contrast, for example, Brazil, where 48 per cent of the income goes to the top 10 per cent of the population, and just 0.8 per cent of total income goes to the bottom 10 per cent, with India, where 25 per cent of the income goes to the top 10 per cent of the population, and 4.1 per cent to the bottom 10 per cent. This way of looking at income distribution indicates one rather obvious point: a large part of any particular distribution is determined by factors inherited from the past – in particular, the stock and distribution of capital (monetary, physical and human). Hence, unless there is very radical action involving asset redistribution (as was taken in Taiwan; Province of China; and the Republic of Korea in the 1950s) or the destruction or flight of human capital (as in Cambodia), we should not expect large changes in income distribution in shortish periods of time.

Statistical investigation shows the importance of asset distribution. For example, education has been estimated to account for 10–20 per cent of observed inequality (Fishlow, 1995); Bourguignon and Morrisson estimate that inequality in land distribution accounts for 17 per cent of income inequality; they also show that the abundance of mineral resources in a country has to be associated with higher inequality (Bourguignon and Morrisson, 1990). Aside from asset redistribution, changes in income distribution depend on changes in the amount and returns to current activities, that is to say, employment of different types of labour, and returns to that employment. The poor are invariably among those with low earnings. Some work for long hours but at very low rates of remuneration, and others have little or no employment. Some combination of low hours of work and low remuneration is normally found in the agriculture sector, in the informal urban economy, and among some unskilled jobs in the formal sector. From this perspective, growth strategies likely to improve the earnings of the poor, and hence improve income distribution, include those which raise returns to agriculture, increase the availability of unskilled work, and extend basic education.

This analysis and cross-country comparisons suggest the following factors are likely to lead to more egalitarian growth:

- Agrarian-focused strategies, especially those also favouring rural industrialization;
- Employment-intensive strategies (export-led, and/or supported by labour-intensive employment schemes);
- High levels of and widespread education;
- Asset redistribution. This is important, not only for the immediate impact on income from assets, but also because it influences the rest of the development strategy in a variety of ways; for example, more equality leads to more widely spread education, and it may, lead to mass markets for labour-intensive consumer goods rather than élite goods. The consequent political economy tends to favour more pro-poor economic decisions;
- Government policies towards structuring the market, so that education, training, and asset accumulation is directed towards deprived groups.

Examples of each of these approaches are presented briefly below:

(a) Agrarian-Focused Strategies

Considering that in the poorest countries a majority of the population works in agriculture, that the rural sector is typically poorer than the urban one, and that the percentage of the labour force in agriculture generally exceeds the share of agriculture in GDP, increasing the productivity of the rural sector should clearly promote more equitable growth. Indeed, Lipton saw 'urban bias' as the major source of poverty in developing countries (Lipton, 1977).

A wide range of strategies would tend to promote agrarian-focused growth. These include reducing macroeconomic biases against agriculture, which almost always arise as countries attempt to promote industrialization; introduce land reform; improve access to extension services, market information and input and output markets for smallholders; develop rural infrastructure; and promote agricultural diversification and the non-farm rural sector (*e.g.* through agro-processing, which has forward and backward linkages, and can employ those with little or no land).

One study estimated that in Pakistan, in the absence of government price interventions, farm incomes from the five major crops would have been 40 per cent higher over the 1983–1987 period (Dorosh and Valdes, 1990). The effect of the price interventions for the five major crops translated into a transfer out of agriculture of

25 per cent of GDP in the 1978–1987 period. In contrast, Taiwan Province of China had relatively moderate macro-bias against agriculture, strong and egalitarian agricultural growth and high levels of rural industrialization (Burmeister *et al.*, 1999).

Most countries that have done well in agriculture have taxed the sector moderately and provided strong support for it. For example, in Malaysia, in the 1970s, total commodity taxation was relatively low at 19 per cent of value of output, while government spending in direct support of agriculture was 10 per cent of the sector's value added. Malaysia had an agricultural growth rate of 5 per cent per year. By contrast, in the same decade, Ghana taxed agricultural commodities by over 60 per cent and spent only 3 per cent of value added on support – its farm output fell more than 1 per cent per year. Similarly, Latin America taxed agriculture much more heavily than East Asia (directly and via an overvalued exchange rate) and the growth in agricultural output and productivity was substantially lower (Schiff and Valdes, 1992).

Rural industrialization also tends to improve income distribution. This is more likely to be dynamic where agricultural output is rising fast, so that there are strong agriculture/non-agriculture linkages; such linkages can also be promoted by government support for rural infrastructure and credit, and are likely to be stronger where land distribution is more egalitarian (Ranis and Stewart, 1987). Decomposition of the factors accounting for the increase in equality in Taiwan Province of China over the 1960s has shown that an important element was rising household incomes from non-agricultural sources (Fei, Ranis and Kuo, 1979).

A study in Bangladesh showed the strong impact of investment in rural infrastructure on rural incomes. A comparison between villages which had benefited from greater provision of infrastructure compared with those that had not, found a one-third increase in average household incomes among the beneficiary villages. Crop income grew by 24 per cent, wage income by 92 per cent, and livestock/fishery income by 78 per cent, all benefiting the poor. Non-farm businesses increased by 17 per cent, which benefited both the non-poor and the poor via improved non-farm employment opportunities (World Bank, 1990).

(b) Employment-intensive Strategies

Quite sharp differences can be observed in the employment-intensity of output increases. For example, the employment elasticity with respect to output growth was estimated at *plus* 0.5 in East and South-East Asia (1971–1992), compared with *minus* 0.5 in Latin America. Within Asia, it was higher in some places (+0.7 in Indonesia), and lower elsewhere (for the 1980s, +0.3 in India, negative in the Philippines). In general, more rapid growth in employment is likely to be associated with more egalitarian income distribution. These differences are partly due to the output mix between and within sectors, and partly to choice of technique. Rapid expansion of labour-intensive exports – observed in many Asian countries – contributes to fast growth in employment. Policies which support this, and also favour more labour- intensive techniques (*e.g.* by not subsidizing capital and by securing more credit for small enterprises) promote employment-intensity (Stewart, 1987). Trade liberalization may increase labour-intensive exports in countries with abundant labour, but only if infrastructure is adequate and the labour has at least minimal education (Wood, 1994). In economies, where the dominant export is minerals or plantation agricultural crops, trade liberalization can worsen income distribution. Employment schemes can also contribute – for example, the Maharashtra Employment Scheme, the very extensive employment schemes in Chile in the early 1980s, and in Bangladesh in the 1970s.

(c) Asset Redistribution

A more egalitarian asset distribution not only contributes to more equality in income distribution directly, but also indirectly by increasing the employment intensity of output in both agriculture and non- agriculture, and strengthening domestic linkages (*i.e.* the demands that agriculture generates for non- agriculture; and the demands that the formal sector generates for the products of the informal sector (Ranis and Stewart, 1987, 1993, 1999)). More equal distribution of land also raises output – Deininger and Squire (1997) show that a difference of one standard deviation in the initial Gini coefficient for land is associated with income gains of 0.5 per cent for the population as a whole, with gains of 1 per cent for the poorest 20 per cent and 0.9 per cent for the poorest 40 per cent.

A comparison between Viet Nam and Bangladesh provides an illustration: in Viet Nam, land distribution is much more egalitarian and there is much less landlessness; Viet Nam has experienced an agricultural growth rate of 5 per cent per annum, compared with a rate of 2 per cent in Bangladesh over the past decade (Ahmed and Goletti, 1998).

Land reform has been very effective in some economies (*e.g.* Taiwan Province of China and the Republic of Korea, and also Egypt in the 1960s), but political obstacles are often severe. In quite a number of countries, even though reforms have been only partially implemented, substantial land redistribution has been achieved. (Powelson, 1984; Lipton, 1993). Moreover, even the more limited reforms generate some improvements in rural income distribution (El-Ghonemy, 1990; Besley and Burgess, 1998). There is a strong case for making land reform a high priority in strategies to improve equity and growth.

With industrialization, land distribution is becoming of lesser aggregate significance, while the distribution of industrial and financial assets is of growing importance. The public ownership of such assets previously represented a way of moderating asset inequality, but this is no longer regarded as a policy option. In highly unequal societies, as in South Africa, more direct policies to tackle asset inequalities are needed. Wealth and inheritance taxation is one option.

The distribution of credit influences accumulation and the distribution of assets. In general, formal-sector credit tends to be biased against the low-income groups because of their lack of collateral, while informal sources are extremely expensive. Surveys of the informal sector generally report that less than one per cent have access to formal sector credit (see, for example, surveys quoted in IADB, 1999; Anderson, 1982; Stewart, Thomas and de Wilde, 1990). The self-employed and employees of micro-enterprises are generally among the lower-income groups; for example, in Latin America it is estimated these enterprises account for 30–40 per cent of low-income earners. New lending mechanisms, such as the group lending procedures of the Grameen Bank, can help to redirect credit to low-income groups.

(d) Education

According to Thompson (1998), "Considering the high payoff from investment in human capital, the unequal distribution of education opportunities is often a more important determinant of the skewed income distribution than is the skewed access to land". Higher (primary and secondary) school enrolment rates tend to be associated with lower inequality. On the basis of cross-country analysis, Bourguignon and Morrisson estimate that a one per cent increase in the share of the labour force having at least secondary education increases the share of income received by the bottom 40 per cent by 6 per cent, and that received by the bottom 60 per cent by 15 per cent (Bourguignon and Morrisson, 1990). One recent study (Behrman, 1993) finds that those with no schooling have a 56 per cent probability of being among the poorest 20 per cent, while those with university training have only a 4 per cent chance. In Brazil, an average 25-year-old in the top decile has an average of 11 years of education, while one in the bottom decile has just two years of education (IADB, 1999).

A study of Latin America in the 1980s found that about a quarter of the inequality among workers' incomes was due to differences in educational levels (Psacharopoulos *et al.*, 1996). Education benefits the rural as well as the urban population – educated farmers are more likely to adopt new technologies and obtain higher returns on land (shown by studies in Malaysia, the Republic of Korea and Thailand) Chou and Lau (1987) show that in Thailand one additional year of schooling adds about 2.5 per cent to farm output. Even in the informal sector, there seem to be high returns to education. Returns were estimated to be as high as 33 per cent for women self-employed in the retail textile sector in Peru, and 14 per cent for post-primary educated men in the service sector (World Bank, 1990).

Educational access and expenditure is often distributed very unevenly. Adult literacy varies from over 80 per cent in East Asia to as little as 13 per cent in Niger, and only 36 per cent in Pakistan. Zambia spent nearly a quarter of its educational budget on tertiary education, with an estimated enrolment rate of 2 per cent in 1980, while Bangladesh spent 8 per cent with an enrolment rate of 3 per cent, and the Republic of Korea also spent 8 per cent, but with a much greater enrolment rate of 48 per cent.

Increasing educational access improves both equity and efficiency, and can make other reforms more effective, such as those aimed at promoting agricultural growth or labour-intensive exports.

(e) Structuring the Market

This describes a set of policies directed towards ensuring that particular deprived groups get favoured access to assets, including education, in a market economy. A range of policies can be used to this effect; for example, employment regulations that require enterprises to employ a certain proportion of a targeted group, such as those introduced as part of the Africanization policies of many newly independent African States. Restrictions can also be directed towards educational institutions (as in the positive discrimination policies in the United States), towards banks, in the distribution of government expenditure, and so on.

The Malaysian New Economic Policy is another example. In Malaysia in 1970, the majority population (the Malays or Bumiputra) were economically the most disadvantaged section of the population, with an average household income 40 per cent less than that of the ethnic Chinese income. A New Economic Policy (NEP) was introduced designed to improve the economic position of the Malays. Targets were defined: 95 per cent of new lands were to be settled on Malays; at least 30 per cent of the equity of all public companies was to be owned by Malays; educational quotas in public institutions were specified in line with population shares; credit policies favoured Malays, with credit allocations and more favourable interest rates.

The policy was a success from many perspectives. The employment share of Malays in manufacturing rose from 26 per cent to 41 per cent between 1967 and 1987, with their share in professional and technical employment rising from 47 per cent to 56 per cent; the share of ownership in public companies rose from 4.3 per cent in 1971 to 19.4 per cent in 1988; their university enrolment share rose from 12 per cent in 1969 to 61.8 per cent in 1988. The income gap between Malays and other groups was almost eliminated. The success of the structured market in Malaysia in meeting its own objectives (narrowing gaps between Malays and others) was achieved without undermining growth, while income distribution improved. Malaysia's growth rate over the period was one of the fastest in the world, at 6.3 per cent per annum during the period 1960–1989, while there was a marked improvement in income distribution, with the share of income of the bottom 40 per cent rising from 11 per cent in 1970 to 14 per cent in 1987, and the share of the top 10 per cent falling from 41 per cent to 35 per cent over the same period.

19.3. The Global Economy and the Decline of Social Reform

Changes in the postwar development of the welfare state became evident during the 1970s, and despite national variations these changes revealed numerous international parallels. In the context of fiscal crises and lower rates of growth, governments introduced shifts in tax structures, caps on public spending, wage and price controls, and "anti-inflation" monetary policies. Further expansion of reforms slowed or stopped, and the principle of the welfare state itself began to come under attack. The general cause of these shifts was not a changed political vision but rather what has come to be called the "new reality," the byword for the internationalization of capital, the coming of the global economy. In this new world market capital became "denationalized," that is, characterized by such a high degree of productive capacity and interlocking trade and investment that it cannot be said to have a national home. This is not to say that the largest and most industrialized nations no longer act in the interests of their "national" capital or that the national state is no longer an important source and means of capital accumulation. It is, however, to say that, once internationalized, capital: 1) requires "freedom" from national controls or intervention; 2) has ultimately no national allegiance; 3) has interests that span the world and far exceed national jurisdictions;

and 4) operates within a world economy and possesses a "global perspective" in which domestic or national markets form only one element and indeed are too small for the productive capacity it possesses. The global economy produces its own demands that are distinct from those of the national economy. It also produces an integrated international arena that is both distinct from and a challenge to bounded and protected national economies, now moribund as such but characteristic of the development of capitalism up to the 1970s.

It follows that the persistence of a national framework in the form of trade barriers, monetary and fiscal regulations and institutions, and a vast number of laws and standards must be overcome. Just as with the formation of the nation-state in the eighteenth and nineteenth centuries, when local and regional barriers to trade had to be eliminated and national standards established in order to provide the possibilities for capital to expand more or less unfettered within a defined geographic territory, so too national standards and barriers to trade now have to be eliminated to allow capital to compete without such restraints in a global economy. We have arrived at the end of the era of the nation-state, and its declining significance is an important factor in the erosion of social reforms. Economic powers are now able to determine the direction of development at the international level, where compromises with national working classes, and indeed national governments, are not so necessary. Throughout the history of the capitalist mode of production there has been an expanding world market. Capitalist classes have always sought to encompass and capitalize not only their domestic markets but also the "external" market. Such expansion, moreover, has been facilitated by a corresponding growth in the size and power of the state, in the public debt, and in financial mechanisms, that is, in international banking and credit facilities. There have also been correlative developments in the nature of the labour supply: the development of capitalism engendered the forcible "freeing" of farmers or peasants from their land, the enslavement of non-Europeans, the purchase of indentured or bonded workers, the export of convicts, and the use of child labour.

To be sure, a world economic system was evolving, but it was coming about through the expansion of nationally defined corporations that were branching out, in competition with other capital formations, to increase their overseas territorial dominions as markets for their commodities, sources of raw materials, and avenues for investment. It was defined by the increasing interdependence of national capitals, but it was still a system of national capitals whose interests were promoted by a national state, and whose expansion gave rise to the modern colonial system, forms of protectionism, and international wars. By World War I, the most industrialized national capitals supported by state military power had divided the world into spheres of influence, but they had also spilled over their national foundations in other ways.

They had formed international cartels, trusts, and banking syndicates for the control of world markets; and after the First World War they had attempted to create international organizations and regulatory agreements to mitigate imperial rivalry (League of Nations), to regulate inter-imperial trade, investment, and capital markets (Bank of International Settlements, BIS), and to help guard against the expansion of socialism (the tripartism of the International Labour Organization, ILO). Despite these transnational beginnings, the world economic order between the two world wars remained dominated by competing national capitals. The Second World War, like the First, represented competition between national capitals raised to the level of the mobilization of the resources of the entire nation. It was, however, to be the last of such wars. The immediate preconditions of globalization lay in World War II, the last major conflict characterized by the expansion of capital in its national form. Prior to that war, capital was largely national in character, and corporate interests, as "national interests," were jealousy guarded at home and abroad by means of standing armed forces, tariff barriers, currency controls, and nationalism and citizenship, among numerous other national laws and policies. The war represented the historical denouement of the contradiction between the continuous impetus of capitalist expansion and its form as the nation-state.

From the point of view of capital, the war's end presented two dilemmas: how to reconstruct capitalism in Europe - but in a way that avoided rebuilding the contradiction between expanding capital and its national political form - and how to contain socialism, a form of property antithetical to the growth of capitalism. The solutions were more often than not combined in the same developments. For the capitalist nations, the period between World War II and the 1970s constituted an interregnum between the age of competing imperial powers and the coming of the global economy. As the pre-eminent capitalist power at the end of the war, the United States was in a position to restructure the postwar world to reflect its own interests. Those interests, however, were in essence the interests of capital as capital, albeit still in national guise. Given this, the U.S. restructuring of the postwar capitalist world marked the beginning of the resolution of the contradiction that had given rise to both world wars. The world as it was - national governments representing national corporations - could grow no further except to encounter more collisions between national forms and with the obvious outcomes - bilateralisms, commercial warfare, national economic problems, and war. Moreover, given new atomic weaponry, another world war between the industrial nations had become unthinkable. So began a period in which the boundaries of international relations were largely prescribed by the claims and exigencies of one imperial power. The "international regime" introduced after the war was a reflection of the international demands of U.S. corporate interests. As the dominant postwar power, hitherto limited in its access to the European and Japanese economies and their colonies or spheres of influence, the United States was intent on opening up new markets to itself. Its efforts were to lay the groundwork for a single world economy of competing capitals.

The U.S.-designed postwar rapprochement saw the creation of political and economic structures and policies that remained dominated by the United States throughout the period. Under U.S. auspices in 1944, the future members of the United Nations (UN) held a conference at Bretton Woods and established the basis of an international monetary system. The delegates created an exchange-rate mechanism by setting the parities of national currencies against the U.S. dollar; and they established, among others, two institutions, the International Monetary Fund

(IMF) and the World Bank (WB). The IMF was intended to regulate international trade balances, but its general function was described as "a sort of Magna Carta for a future world economic order." The World Bank was designed to manage an international fund for economic development. Alongside these mechanisms, the General Agreement on Tariffs and Trade (GATT, established in 1947) would provide the institutional means for a negotiated removal of all national barriers to world trade and to create universal regulations for increasingly freer commerce.

Although these economic institutions and agencies were of primary importance after the war, the United Nations, dominated by the United States, was formed as the beginning of supranational quasi-government, as the political foundation for the new internationalism. Among other things, the UN took upon itself powers to oversee the long process of decolonization (one of the purposes of peace-keeping operations), which in most instances aided U.S. penetration into the former European colonies and advanced the geographic expansion of the capitalist mode of production and the destruction of remaining precapitalist modes. The UN was also employed to contain the expansion of socialism (Korea, the Congo, Greece, Indonesia), to promote the principles of liberal democracy and the rights of private property as a political system consistent with the advance of capitalism, and to establish international laws grounded in private rights for the "world community."

Although the United Nations appeared to be the embodiment of a genuinely democratic and multilateral world in the making, it was dominated by an executive (the Security Council), whose members (with the exception of the U.S.S.R.) were largely beholden to the United States. The UN was, moreover, financially dependent on the United States, as were most of its member nations, making it largely an instrument to make the world over in the image of America. This quintessentially capitalist nation and the principles that it stood for were to be the model for the new world in the making, and U.S. dominance in supranational agencies and institutions remains to this day. The liberal democracy promoted and established in the former Axis regimes and the decolonized world was the Trojan Horse of the new order - it held no intrinsic national character. Its principles were the principles of corporate private property, and the political system would amount to the periodic choice of different policies within this property regime; and as a result those principles would ultimately overreach national barriers. The assertion of national interests was possible, and they were promoted, to be sure, but in the final analysis national goals would prove contrary to the embedded political principles.

Besides liberalized world trade, the containment of socialism, and the furtherance of liberal political and legal principles, another pillar of the new internationalism was the aggregation of national markets. As early as the late 1940s the United States encouraged the European countries to see their future as the "United States of Europe" (a larger market for the postwar expansion of capital), and to this end they entered into the Organization for European Economic Co-operation (OEEC), whose main function was to promote intra-European free trade. The Treaty of Rome (1957) soon followed, establishing the progressive growth of a European economic community. But Europe was not alone; by the 1960s, almost the entire world was part of one or another customs union, common market, or free-trade area.

Although engineered by the United States to suit its economic needs and prevent the imperial conflicts of the past, this rapprochement contained contradictions that were to frame the period of its existence. The most obvious concerned the ascendency of U.S. military and economic power over the industrial nations of Europe and Japan. In effect, the United States represented one configuration of national capital interests, albeit the most powerful, in a postwar world in which other national capitals and their states were far from moribund, and in which their economic reconstruction was necessary to the success of the new world order. A second contradiction involved the persistence of national interests in the face of the developing "international regime." With no realistic alternative to an international future, national capitals were confronted with the dilemma of defending national interests while embracing the internationalism necessary for expansion. Another contradiction came with the struggle between capitalism and socialism as the two main postwar roads to the future.

Throughout Europe the spread of socialism after the war as a promise of workers' power and a form of property antithetical to capitalism was pervasive. To ensure the future of capitalism, socialism had to be contained, and to this end the U.S. administration took up several initiatives. One comprised a massive financing of capitalist reconstruction on strictly U.S. terms (UNRRA- the United Nations Relief and Rehabilitation Administration, set up in 1943 to provide material relief for liberated nations - and the European Recovery Programme, or Marshall Plan, initiated in 1947), which involved a substantial degree of state intervention in both capital and working-class reproduction. Another involved the U.S. promotion of Keynesianism as a means of reconstruction that gave the working class varying degrees of protection from the effects of capitalism and involvement in its reconstruction. To affirm the U.S. commitment to this reconstruction of capitalism, a military alliance, NATO, was created in 1949 with a permanent U.S. commander-in-chief ensuring that country's military dominance over Europe. In addition the postwar period saw interference in European elections, political parties, government policies, and trade union movements. Although socialism as state capitalism was ultimately abandoned in the 1980s and 1990s, the postwar period saw the consolidation of Soviet hegemony over Eastern Europe, the victory of the Chinese People's Liberation Army, and numerous wars of liberation in former colonial countries. All of these events suggested that the socialist road might indeed succeed, a possibility that was always implied in the co-operative military and economic arrangements of the interregnum.

The same problems of rebuilding and consolidating capitalism, containing socialism, and ending national barriers to accumulation also presented themselves in the European colonial empires and spheres of influence. Within one decade of the end of World War II, struggles of national liberation arose in most of the colonial world: in Vietnam, China, Malaysia, Dutch East Indies, Korea, the Philippines, and India; in Algeria, Egypt, and Kenya; and in the Middle East. For U.S. capital, and capital in general, the socialist and nationalist movements of liberation had to be fought or co-opted, and the colonial systems had to be dismantled and opened to international trade. While military means were employed in many of these cases, greatly supported by the United States, the chief means of disengaging the colonial link and establishing the foundations of a capitalist economy were United Nations peacekeeping (both military and diplomatic operations) and the installation in the colonies of liberal-democratic or dictatorial regimes as required. The colonial systems could no longer be tolerated because their very nature favoured restricted trade relations with the metropole, while liberal democracy (or, in many cases, authoritarian systems) opened the door to foreign investment and embraced the principles that would ultimately supersede the structure of the nation-state.

To bring an end to colonialism and to ensure that the newly independent countries remained within the orbit of international capital accumulation, the preferred solution appeared to be a form of political independence embracing the formalities of liberal democracy and constitutionally entrenched principles of private property. Neo-colonialism - the subordination of former colonial countries mainly by economic means and principles - was established as the new relation between the industrial metropoles and the "lesser developed" nations. Transformed from colony to nominally independent nation-state, from subordination to one industrial metropole to compliance with the new international regime of corporate private property, most of the former colonies were, under the guise of political freedom, opened to international trade. A political shell of independence without the "content" of an indigenous capital formation obscured the real content, which was in essence the multinational corporation of the industrial world. From the point of view of capital, both the war and its aftermath represented the restoration of the conditions for renewed accumulation. The war brought an end to economic depression, with state intervention in labour supply (regulated wages, restricted union activity, and even slave-labour in Germany and Japan) and in corporate assistance (superprofits and tax concessions for war production); government mobilization of public and private capital; and the destruction of inventories and new products in the war effort. After the war, the basis for renewed accumulation became the reconstruction of the devastated economies, again promoted by state-levered capital for the corporate sector by means of deficit financing, increased taxes, debased money, and, very importantly, U.S. "aid," particularly in the form of UNRRA and the Marshall Plan. Moreover, new markets and sources of raw materials emerged as many of the former colonial territories were opened to international capital.

Within a decade after 1945, the main elements of a framework of a supranational regime of accumulation, albeit dominated by the United States, had been established. There were the principles of liberal democracy ensconced in the United Nations, Marshall Aid for reconstruction, the IMF, World Bank, and GATT to regulate and liberalize world trade, incipient common markets promoted everywhere, and U.S. military control and influence throughout the world. All these facets of the postwar world order were progressively realized and operative by the end of the 1960s. A foundation had been laid for the regulation of multinational corporations that in principle transcended bilateral or multilateral relations between nation-states. This global regulatory regime was an essential precondition, a necessary and complementary part of the transnational corporation and the process of globalization.

Extranational economic expansion was eroding the power and influence of the national state. Corporate trade and investment overseas were growing dramatically as a percentage of total operations, and the multinational corporations (MNCs) had begun shifting profits and operations to places where the advantages in taxation, wages, or state support were greatest. Increasingly, they assumed the powers of sovereign entities, ignoring and interfering in the interests of nation-states. This expansion of the MNCs added international pressure to remove national barriers and level the costs of corporate activities. The rise of foreign currency and international capital markets, moreover, meant not only that these economic functions were increasingly outside the control of national regulation, but also that they increasingly set the limits, as independent markets, on national policies with respect to credit systems, money supply, exchange and interest rates, debt-management, investment policies, and taxation. The growing erosion of the role and powers of the national state was also visible in the rise of supranational accords or treaties, multilateral agreements and organizations that all in their own way began to restrict national sovereignty. Between 1945 and the early 1970s, international government organizations (IGOs) and international non-governmental organizations (INGOs) broadened their powers and increased by about threefold in number, such that by the end of the 1970s there was hardly an aspect of national life in industrial countries that was not addressed in some way by a supranational body. Military multilateralism in the shape of NATO, the Warsaw Pact, the United Nations, and other regional pacts took the place of the independent national exercise of armed force outside national borders. National policies on international trade, scientific and technical standards for goods and services, transportation and communication, employment and health standards, and pollution were all increasingly subject to modification and reformulation to bring them into accord with internationally set rules and regulations.

The decline in powers of the national state was also reflected in the worldwide increase in preferential trading groups. Although the most important and earliest was the European Economic Community (EEC) in 1958, there was not a single region in the world, including the centrally planned economies, that had not attempted to create a common market, customs union or free-trade area between the 1950s and the mid-1970s. Despite certain political resistance and at times faltering success, economic interdependence at the global and regional levels had everywhere profoundly circumscribed national sovereignty. From 1945 to the early 1970s, capitalism developed in a two-tiered manner that reflected the nature of the postwar rapprochement: the state-managed reconstruction of capitalism took place within the nation-state, while at the same time the framework for international free trade was undermining the national economy, the rationale of the nation-state. The contradiction between the national form of capital and the intrinsic drive of capital to overcome this shell was moving towards its resolution with the erosion of this rationale. Given the expanding supranational economic framework, the size and power of MNCs, and the actual global development of capitalism, the only resolution remaining to the contradiction was the dismantling of the national shell. By the early 1970s, the national relations of production were increasingly seen by transnational corporations (TNCs) as mere fetters to intercorporate relations at the global level.

Throughout this postwar period, and corresponding to the growth of MNCs and international government organizations, there arose international strata of bureaucrats, technocrats, academics, and corporate managers, owners, and advisors. They managed the agencies and organs of supranational governance, and the highest placed also met regularly in certain forums for the express purpose of discussing global trends and policies for the advance of certain interests. Besides the more "public" venues, such as UN agencies, the Organization for Economic Cooperation and Development (OECD), and GATT, private venues such as the Trilateral Commission, the Bilderberg conferences, the Club of Rome, and the so-called think-tanks such as the Mont Pelerin Society and the Heritage Foundation were all employed for these purposes. While not coherent as a national government, here nevertheless was a growing matrix of "public" and private organizations for decisionmaking and policy-formation at the global level that became increasingly influential as the demands for a global economy increased, and that formed the meeting-ground for transnational strata occupied with the advancement of a global economy.

It is perhaps not possible to say when the global economy became a reality as distinct from interrelated national economies, but certainly by the mid-1970s the end of the interregnum could be seen in two significant developments. One was the extensive reduction in barriers to world trade and production: the productivity of the advanced Fordist mode of production possessed the capacity for world supply, and the postwar "enabling framework" had gone far to erode national restrictions. The other was the loss of U.S. national paramountcy, signaled by several events. Perhaps the most important was the suspension of the convertibility of the dollar into gold in 1971, which ended the fixed exchange rate mechanism that had so benefited U.S. expansion. Another was the waning of the rate of U.S. direct investment and exports to Europe by the early 1970s, while European and Japanese products and capital began to claim ever greater shares of the U.S. market. Also about this time the United States had begun to lose some of its edge in technological superiority to Germany and Japan. On the surface, the United States appeared to be yielding to these rival economic powers; in effect, it was succumbing to the internationalization of capital - built on a foundation that Washington itself had forced on the world. Americanization was becoming tempered by internationalization; the premise was becoming absorbed by its conclusion.

There were other harbingers of the end of the interregnum or the approaching global economy, and all were consequences of two decades of unprecedented capital accumulation. One was the near completion of decolonization and the consequent "opening up" of the entire globe (with the exception of the "socialist" countries) to capitalization. The world was now in effect subjected to the capitalist mode of production. Another precursor was the transformation of increasing numbers of multinational corporations into TNCs, defined by the expanding internationalization of their production and decline of a meaningful home base or domestic market. Related to this development was the rising importance of international foreign exchange and capital markets and of supranational agreements and organizations, and the consequent increasingly obvious erosion of national identity and sovereignty, not to mention the globalization of culture. Ironically, at the culmination of this interregnum, and before the transformation of the nation-state was about to begin in earnest, the voice of the non-corporate sector began to be heard. In the industrialized nations, a citizenry that had come to believe that their right to vote gave them power over policy began to demand and expect a progressive extension of the welfare state - demands characterized by the corporate sector as the "ungovernability of democracy." In the decolonized nations, leaders labouring under the belief that independence would bring freedom, democracy, and material betterment and not a new form of colonization made demands for an end to unequal exchange, a new international economic order. More successfully, a group of oil-producing nations formed a loose coalition (OPEC) and began to exact a somewhat fairer share of oil revenues, albeit mainly for their own elites, from the oil cartel. By the end of the 1970s, as the "new reality" began to set in, these voices were silenced or compromised, or had found other forms of expression.

By about 1970, the key institutional preconditions for a global economy were in place; they had become a force of production in the encouragement of global economic interdependence and helped foster new means of production. Nevertheless, production was still largely national, although much trade and commerce had turned international. The movement to a global economy in which the circuit of capital itself was global and national integrity weak and in decline had not yet become manifest. About this time a dramatic change in the technology of production, distribution, and circulation began to be realized. These were changes that could not possibly be contained within the framework of the nation-state.

The trends towards the internationalization of capital during the interregnum continued and intensified in the 1970s and thereafter. Commodity trade expanded, though it was now in the shape of world commodities in a world market: by 1980 the sales of many TNCs in "foreign markets" increased to more than one-half of total sales; and the imports and exports of all the industrial nations, except Japan, increased as a percentage of the GNP. A system of highly integrated world trade was an irreversible fact by the end of the 1970s, confirmed and hastened by the new means of transportation and communications, whose increased productivity were transforming the worldwide distribution of products and hence the global conditions for valorization. Also, with the coming of computer-based systems, capitalist production became less constrained by large-scale, long-term investments; and with the continuing diminution of national barriers it could increasingly be "sited" in regions or countries with the most favourable conditions for accumulation (for instance, proximity to markets, types of labour force, and state concessions). This locational flexibility of productive capital became an important factor in the development of a changing international division of labour, and the 1970s saw a geographic restructuring of production leading to a rapid rise in the world share of manufactured exports from developing countries. Along with this came a growth in the Eurocurrency markets, to such a degree that by the end of the 1970s the world's financial markets were close to becoming a single integrated market.

20

Overview of Nigeria Economy

20.1 Overview of Nigeria Economy

As of 2001, the most conspicuous fact about Nigeria's economy is that the corruption and mismanagement of its post-colonial governments has prevented the channeling of the country's abundant natural and human resources–especially its wealth in crude oil–into lasting improvements in infrastructure and the construction of a sound base for self-sustaining economic development. Thus, despite its abundant resources, Nigeria is poorer today than it was at independence in 1960. Still one of the less developed and poorer countries of the world, it has the potential to become a major economic power if the leaders resolve to learn from past mistakes and to harness the country's rich natural and human resources for a productive and sustained effort to promote economic development.

Before the country was colonized by Britain, during the second half of the 19th century, the various nationality groups that currently make up Nigeria were largely an agricultural people. They were food self-sufficient and produced a variety of commodities that were exported overseas. British colonial administrators amalgamated (joined together) the nationality groups in 1914 into a larger economy for exploitation for the benefit of British industrial classes. Under colonial rule, Nigeria remained an agricultural country, exporting raw materials to Britain and importing from it finished goods. Therein lay the origins of the dependence of Nigerian economy on commodity markets of the industrialized Western world for its foreign exchange. While the industrialization of the country was discouraged, rudimentary foundations for a modern Nigerian economy, however, were laid. Colonial economic policies shaped future independent Nigeria's economy, particularly in marketing, labor supply, and investment. The process of colonial rule and formal economic exploitation ended in 1960 but left Nigeria a relatively strong but undiversified economy. Thereafter, Nigerians were poised to remedy this defect and to build a self-sustaining Nigerian economy comprising agricultural, industrial, and service sectors.

From independence in 1960, the state took up the direction and planning of economic growth and development. Education was progressively expanded at all levels to reduce the rate of illiteracy and to provide the requisite skills and labour force for development. Infrastructure of roads and communication networks were constructed far beyond what was inherited from colonial rule. Hydroelectric dams were built to generate electricity. Secondary industries and automobile assembly plants were established to create more employment opportunities. Because of the paucity (small number) of indigenous (native or local) private capital, these activities were undertaken and financed by the government, often with foreign assistance from such countries as Britain and the United States. Foreign oil companies, such as Shell-BP, Exxon-Mobil, Chevron, Agip, and Texaco, operate in partnership with the government in the oil sector, the mainstay of Nigeria's economy. The capital-intensive oil sector provides 95 per cent of Nigeria's foreign exchange earnings and about 65 per cent of its budgetary revenues.

Because the established, government-owned industries and businesses were often inefficient and corrupt, productivity was low at best. In particular, mismanagement and corruption were endemic (characteristic of) in the successive governments and throughout the nation. However, the gravest problem was caused by the government's decision to stress the industrial sector above all others. Caught in a web of competing demands for scarce resources, the officials took the path of rapid, large-scale industrialization at the expense of the agricultural sector, as well as light manufacturing. They directed the bulk of investment capital towards the promotion of what Western advisers captioned "industrial take off." This decision to abandon the known–agriculture–for the unknown–rapid large-scale industrialization–was a fundamental error. The capital and the skill needed for rapid, large-scale industrialization were not sufficiently available. Thus, an unskilled workforce and insufficient funds severely handicapped the industrial sector. Also, Nigeria's neglect of the agricultural sector aggravated already problematic food shortages. Nigeria had raised enough food to meet domestic needs during its colonial period and in the decade following independence. However, it experienced food shortages in the 1970s and 1980s, which necessitated the importation of food from foreign countries. Among the imports were palm oil (from Malaysia), of which Nigeria had been the world's largest producer and exporter, and rice (from the United States) which was considered less nutritious than Nigerian brown rice. Once Africa's largest poultry producer, Nigeria lost that status because of inefficient corn production and a ban on the importation of corn. Furthermore, it is no longer a major exporter of cocoa, peanuts, and rubber.

Several forces compounded the problems of the agricultural sector. The migration of labor from the rural areas to the urban centers reduced the traditional agricultural labor force. Ecological constraints such as poor soil, erosion, drought, and the absence of agricultural research added to the problem. Other constraints on agricultural production include the use of antiquated technology due to a lack of capital, the low status given to agriculture in the education of the youth, inefficient marketing, an inadequate transportation infrastructure, lack of refrigeration, trade restrictions, under-investment due to unavailability of credit, low prices, and unstable pricing policies which resulted in farmers literally subsidizing urban dwellers and other sectors of the economy. In addition to these handicaps, import constraints limit the availability of many agricultural and food-processing plants. In general, land tenure discourages long-term investment in technology and modern production techniques.

The problem of food shortages and imports was addressed in the late 1970s and early 1980s. In the late 1970s the military government of Olusegun Obasanjo embarked upon "Operation Feed the Nation." His civilian successor, President Shehu Shagari, continued the program as the "Green Revolution." Both programs encouraged Nigerians to grow more food, and urged unemployed urban dwellers to return to the rural areas to grow food crops. The government provided farmers with fertilizers and loans from the World Bank. The food situation has stabilized, although Nigeria still imports food. A related problem which has not been completely resolved is the pollution of water in the Delta region and Ogoniland by oil companies. Water pollution disrupts farming efforts and has been a source of friction between farmers on one side and the national government and the oil companies on the other.

The oil boom which Nigeria experienced in the 1970s helped the nation to recover rapidly from its civil war and at the same time gave great impetus to the government's program of rapid industrialization. Many manufacturing industries sprang up and the economy experienced a rapid growth of about 8 per cent per year that made Nigeria, by 1980, the largest economy in Africa. The growth, however, was not sustained. The new oil wealth did little to reverse widespread poverty and the collapse of even basic infrastructure and social services. The iron and steel industry, started with the help of the Soviet Union, still has not achieved a satisfactory level of production. The oil boom also provoked a shortage of labor in the agricultural sector as members of the rural workforce migrated to jobs in the urban construction boom and a growing informal sector. When the price of crude oil fell and corruption and mismanagement still prevailed at all levels, the economy became severely depressed. The urban unemployment rate rose to 28 per cent in 1992, and crime also increased as 31.4 per cent of the population lived below the poverty line.

Nigeria's debts mounted as administrators engaged in external borrowing and subsidized food and rice imports and gasoline prices. In the 1980s, economic realities forced Ibrahim Babangida's military regime to negotiate a loan with the World Bank and to reschedule Nigeria's external debts. His regime undertook an economic structural adjustment program (ESAP) to reduce Nigeria's dependence on oil and to create a basis for sustainable non-inflationary growth. However, external borrowing to shore up the economy created more problems than it alleviated. Much of the borrowed money never reached Nigeria. The portion that reached the country often went towards abandoned or nonperforming public sector projects. External loans escalated Nigeria's debts to US$30 billion during the

Babangida regime and consumed external earnings in debt servicing. Similarly, the ESAP prescribed by the International Monetary Fund (IMF) failed to advance the economy, and aggravated the problems of inflation and unemployment. It caused a reduction of state spending on education and health care. Continuing political instability due to Babangida's annulment of the presidential election results in June 1993 and the subsequent authoritarian rule of Sani Abacha (1993 to 1998) made the general economic situation worse. The gross corruption by the Abacha regime and its violations of people's fundamental rights turned Nigeria into an international pariah for 6 years, and thus discouraged foreign investment in the economy. Many industries and manufacturing companies could not obtain raw materials and closed down. Others operated under severe handicaps, including rampant power outages and refined petroleum scarcity. Not enough had been done in the years of plenty to diversify the economy or to sustain the development. Military coups (military overthrow of civilian governments) and political instability worsened the situation.

There was considerable optimism in May 1999 when Oluseguan Obasanjo became Nigeria's civilian president. Many hoped that he would lift Nigeria from the verge of economic bankruptcy. One of Obasanjo's objectives to that purpose was to secure debt relief from Nigeria's foreign creditors. However, these creditors insisted that Nigeria's wealth of untapped resources provided the means for the country to pay off its debts, and refused to cancel its debts of US$30 billion. In spite of some opposition, Obasanjo embarked upon a program of privatizing some parastatals in order to reduce corruption, promote efficiency, and raise productivity. He introduced an anti-corruption bill which passed through the legislature, and recovered some of the revenues that had been stolen from the country and deposited in Western banks. The inflation rate, which was estimated at 12.5 per cent at the start of his administration, was estimated at 6.6 per cent in 2000. Significant exports of liquified natural gas started in 1999, and increased crude oil prices in 2000 provided his administration with additional revenues. So far, however, he has been unable to bring about economic recovery. Industrial capacity utilization appears to have diminished. Worse still, infrastructural facilities, including the National Electric Power Authority (NEPA), continue to be in a state of disrepair. Expected massive inflow of foreign investment, on which the government had hinged its economic revival program, failed to materialize. This is in part because of the high cost of doing business in Nigeria and a lack of transparency in economic decision-making in the country. In addition to these realities, the unemployment situation in the country remains unchanged months after the restoration of civilian government. In fact, it has worsened among university graduates and ranged from 30 to 40 per cent in 2000. Political uncertainties due to ethnic and religious conflicts between Muslims and Christians, and constant feuding between the president and the legislators aggravate the economic climate. Widespread armed robbery and a crime syndicate known locally as 419 prey on foreign nationals, further hindering foreign investment and tourism. The country's economy needs the collective efforts of the president and the National Assembly as well as more definite measures to address its ills in order to foster its recovery and growth. Currently, funds available to the government are insufficient to meet the needs of all sectors of the economy at once. External investors can contribute through long-term investment and **joint ventures** in Nigeria's large national market. Crude oil, the price of which rose sharply recently, remains a very considerable asset. Properly managed, it could provide a solid platform for more sustained Nigerian development and prosperity in the 21st century and beyond. Oil is one of the most heavily traded commodities in the world. Fluctuating prices have important effects for oil producers/exporters and the many countries that remain dependent on oil as a key input in their energy, manufacturing and service industries. The demand for oil are:

1. **Cyclical demand:** There is a strong link between the demand for oil and the rate of global economic growth because oil is an essential input into many industries – when the economy is expanding, the demand for oil rises. The best recent example of this is the growth of the Chinese economy. Fast growth of national output in energy-intensive sectors has led to a surge in demand for crude oil into the Chinese economy.
2. **Prices of substitutes:** Demand for crude oil affected by the relative prices of oil substitutes (*e.g.* the market price of gas). If, in the longer term, reliable and relatively cheaper substitutes for oil can be developed, then we might expect to see a shift in demand away from crude oil towards the emerging substitutes. The high price of oil during 2004-2006 seems to have led to a rise in research and development into non-oil substitutes. These can take several years to come through to affect the market for energy.
3. **Changes in climate** – *e.g.* affecting the demand for heating oil. It is often said that if the winter in North America is fierce, then the price of crude rises as the USA and Canadian economies raise their demand for oil to fuel household heating systems and workplaces.
4. **Market speculation:** There is always a **speculative demand** for oil (*i.e.* purchasers hoping for a rise in prices on world markets). Indeed one of the features of the most recent spike in oil prices has been the high level of demand by hedge funds and other investors pouring into the **international petroleum exchanges** to buy up any surplus oil futures contracts. They hope that by the time the contracts are ready to be fulfilled, they will have made a large profit. Speculation involves risk, prices can do down as well as up.

Who are the main consumers of oil? Nearly two thirds of global crude oil production is consumed by the leading industrialised nations – *i.e.* the nations that make up the Organisation of Economic Cooperation and Development. But a rising share of oil demand is coming from the emerging market economies including China, Brazil, Russia and India.

When we consider the global supply of oil we need to make a distinction between short-term and longer-term supply to the international markets. The short run supply curve is normally drawn on the basis of a given state of production technology and fixed use of capital inputs (*i.e.* the oil industry is supplying from a known level of oil reserves and a given stock of capital machinery used to extract that oil). There is inevitably a short-run limit on daily oil supply and, as production gets close to capacity limits, so the short run supply of oil becomes more inelastic.

The World's Largest Consumers of Oil

	Consumption of Oil in 2005 Thousand Barrels Daily	*Share of Total Consumption %*
USA	20655	24.6%
China	6988	8.5%
Japan	5360	6.4%
Russian Federation	2753	3.4%
Germany	2586	3.2%
India	2485	3.0%
South Korea	2308	2.7%
Canada	2241	2.6%
France	1961	2.4%
Mexico	1978	2.3%
Saudi Arabia	1891	2.3%
Italy	1809	2.2%
Brazil	1819	2.2%
United Kingdom	1790	2.2%
Spain	1618	2.1%

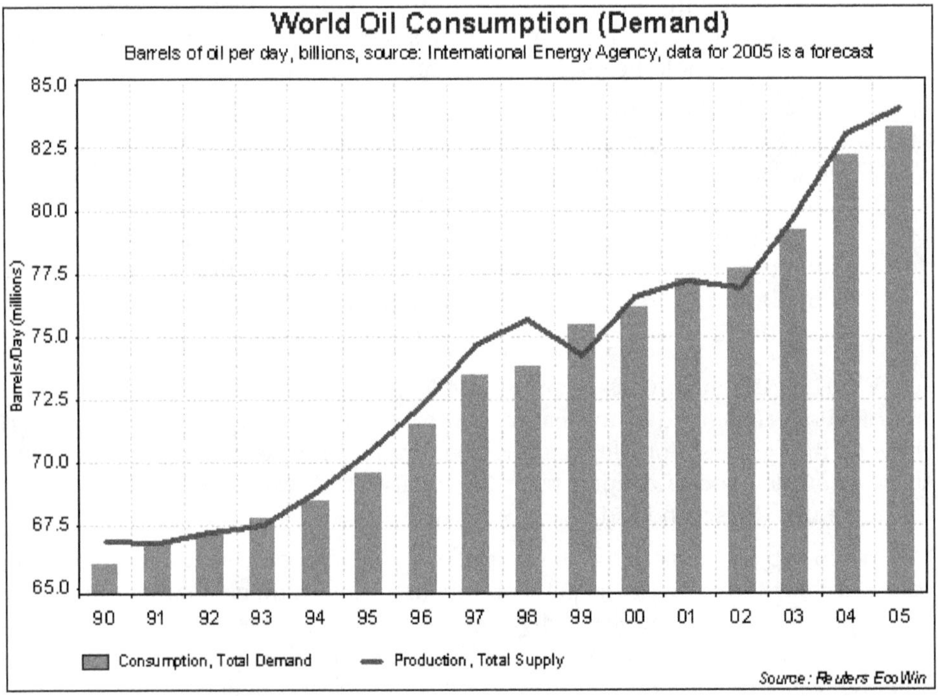

One possible way of modelling this is to assume the market supply curve for oil is non-linear (shown in the left hand diagram below). An alternative is to suggest that more oil can be supplied elastically at a fairly constant price until the capacity limit is reached, when the short run supply curve becomes vertical. In short, the short-run supply of crude oil is affected by a series of different factors:

1. **Profit motive:** The production decisions of OPEC and Non-OPEC countries (see revision notes on OPEC below).
2. **Spare capacity:** The level of spare production capacity in the oil sector
3. **Stocks:** The current level of crude oil stocks (inventories) available for immediate supply from the major oil refineries – *i.e.* a high level of stocks means that extra oil supplies can be released onto the market quickly when demand fluctuates.
4. **External shocks:** The effects of production shocks (*e.g.* loss of output from rig closures or disruption of oil supplies due to war and terrorist attacks). Taking a longer-term perspective, the long run world oil supply is linked to:
    a. **Reserves:** Depletion of proven oil reserves – the faster that demand grows, the quicker the expected rate of depletion.
    b. **Exploration:** Investment spending on exploring, identifying and then exploiting new oil reserves. When oil prices are rising and are expected to stay strong for the foreseeable future, it makes financial sense to invest more resources in exploring for new reserves, even though these may not come on stream for some years.

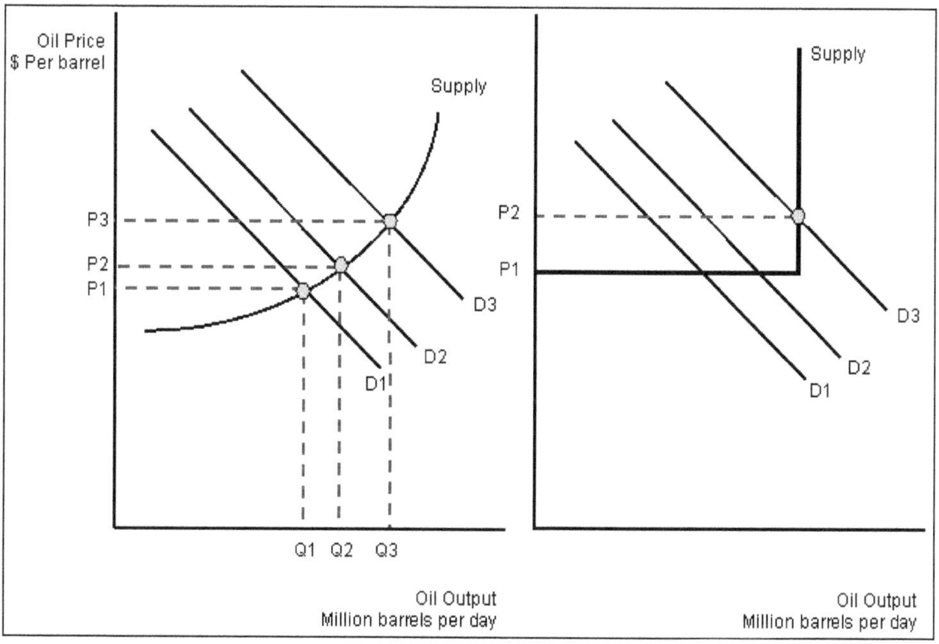

c. **Technology:** Technological change in oil extraction (which affects the costs of extraction and the profitability of extracting and then refining the oil).

Higher oil demand matched against an inelastic short run supply of oil invariably drives market prices higher – this is shown in the diagram below. An increase in demand causes a fall in oil stocks at the major international refineries and pushes prices higher. This acts as a signal to suppliers to expand production. However there are time lags between a change in price and extra supplies coming on stream. The demand for oil is also price inelastic. This combination of an inelastic demand and supply helps to explain some of the volatility in world oil prices.

20.2. The Role and Impact of the OPEC Cartel

The **Organization of Petroleum Exporting Countries** (OPEC) accounts for around 40% of current world supply. This gives OPEC a pivotal influence in shaping the direction of oil prices – but only when the cartel acts together to control production and balance supply and demand in the international market. Non-OPEC countries account for the largest portion of total supply. Oil is produced in nearly every corner of the world, and nearly every region has been expanding oil production in the last decade. This includes Europe, where Norwegian oil companies are achieving a rapid increase in oil extraction and also Russia now one of the world's largest oil suppliers.

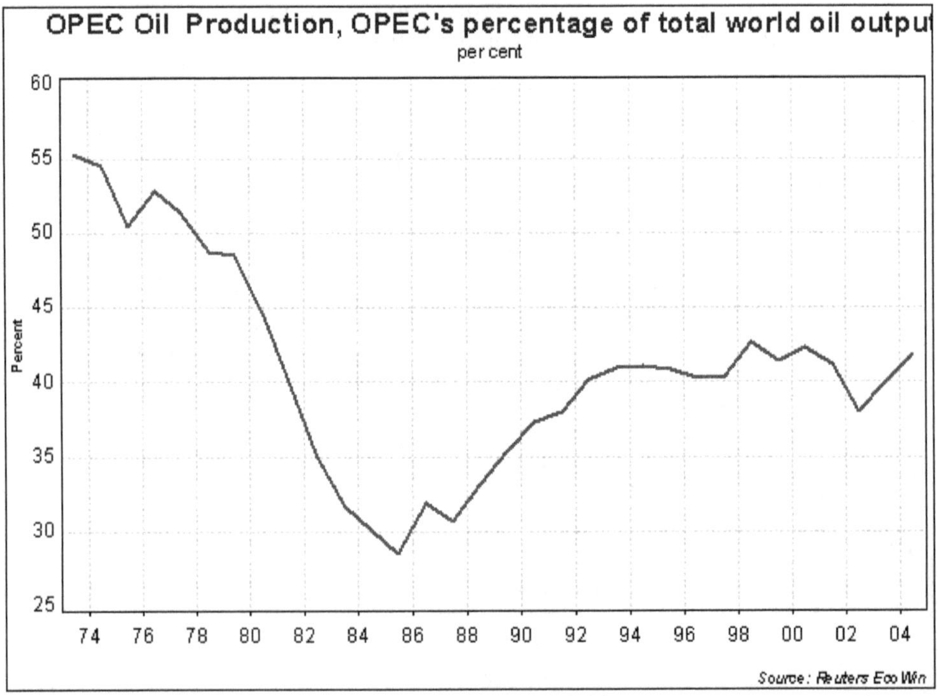

World Oil Production

	Production on Oil Thousand barrels daily	*Output as a Share of World Total*
Saudi Arabia	11035	13.5%
Russian Federation	9551	12.1%
USA	6830	8.0%
Iran	4049	5.1%
Mexico	3759	4.8%
China	3627	4.6%
Venezuela	3007	4.0%
Canada	3047	3.7%
Norway	2969	3.5%
Kuwait	2643	3.3%
United Arab Emirates	2751	3.3%
Nigeria	2580	3.2%
Iraq	1820	2.3%
Algeria	2015	2.2%
Brazil	1718	2.2%
United Kingdom	1808	2.2%

Source: IFPRI Report, 2014.

	Total Production 000 Barrels Daily	*Output as a Percentage of Total World Output*
Total World Oil Production in 2005	81088	100.0%
Of whichOPEC countries	33836	41.7%
Non-OPEC 35408	43.4%	
Former Soviet Union	11844	14.8%

Source: IFPRI Report, 2014.

OPEC sets **quotas** for how much crude oil they want to produce with the aim of stabilising the price at a target level. There are always major doubts about OPEC's ability to keep to output limits. Basically, OPEC acts as the swing producer in the world oil market. It controls that part of the world supply curve which is easiest to change and if it wants to keep oil prices high, then it can keep tight control on short run production so that supply does not run too far ahead of demand. OPEC has to tread a fine line, because if prices remain too high for a long period, then oil consumers have a clear incentive to look for alternative sources of energy or other non-oil substitutes in production.

Crude oil has many uses in many different markets and industries. So changes in the global price of oil inevitably have an effect on the microeconomics of particular sectors of the economy. The main uses for crude oil are as follows:

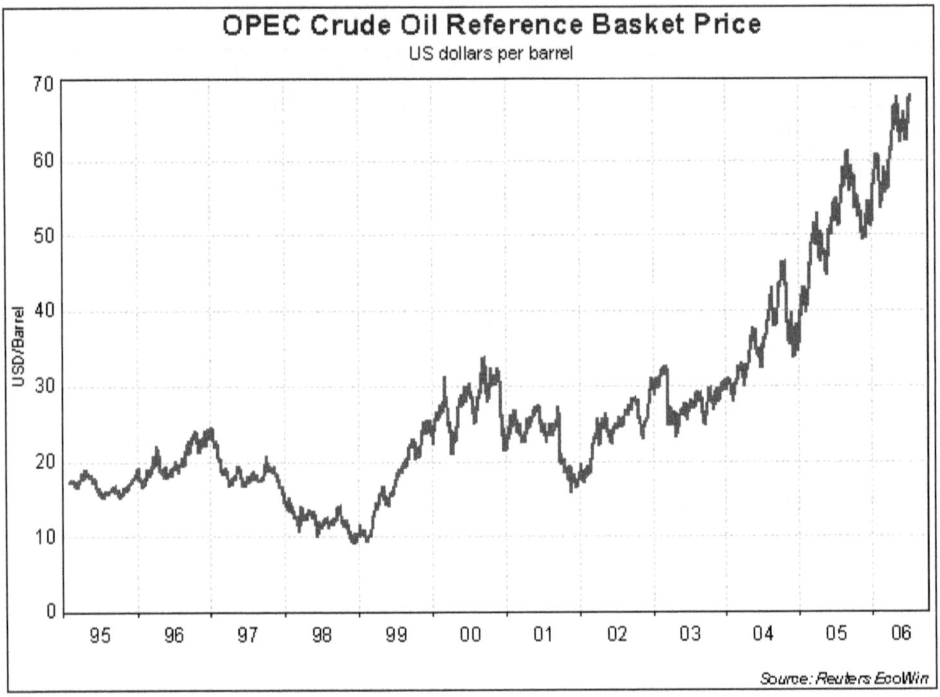

1. Gasoline: motor spirit/petrol
2. Middle Distillates:
    a. Diesel - vehicles and other motors/engines
    b. Jet fuel
    c. Kerosene – cooking/heating
    d. Heating Oil
    e. Fuel Oil: boiler fuel for industry, power and shipping
    6. Other: lubricants, bitumen etc.

After a long period of relatively low oil prices, in the last few years, the world economy has had to come to terms with the prospect that the era of cheap oil is now over. This affects many industries in the UK economy and has direct and indirect effects on consumers.

For those industries that use oil as a key input into their production process, then a rising price acts as a supply-side shock – leading to higher input costs *i.e.* a rise in their variable costs of production. The more an industry relies on oil, the bigger will be the impact of a rise in oil prices on its costs and profitability, and hence the bigger the fall in its production is likely to be in the long run.

The increase in costs causes a profit maximising firm to increase price and reduce the equilibrium level of output. The extent to which a business is able to pass on an increase in costs depends on the price elasticity of demand for their products. If demand is price inelastic, then the supplier may choose to pass on some or all

of any rise in variable costs to the consumer of the final product. For example, a controversial issue has been the decision by many (although not all) of the airlines to increase their fuel surcharges to customers.

For consumers, higher oil prices has led directly to more expensive fuel at the pumps, higher gas and electricity bills and a reduction in their real incomes.

Although oil and gas prices have been very high, so far we have not seen a dramatic rise in inflation – other factors have helped to keep inflation under control.

Nigeria's economy is struggling to leverage the country's vast wealth in fossil fuels in order to displace the crushing poverty that affects about 57 per cent of its population. Economists refer to the coexistence of vast wealth in natural resources and extreme personal poverty in developing countries like Nigeria as the "resource curse". Although "resource curse" is more widely understood to mean an abundance of natural resources which fuels official corruption resulting in a violent competition for the resource by the citizens of the nation. Nigeria's exports of oil and natural gas–at a time of peak prices–have enabled the country to post merchandise trade and current account surpluses in recent years. Reportedly, 80 per cent of Nigeria's energy revenues flow to the government, 16 per cent cover operational costs, and the remaining 4 per cent go to investors. However, the World Bank has estimated that as a result of corruption 80 per cent of energy revenues benefit only 1 per cent of the population. In 2005, Nigeria achieved a milestone agreement with the Paris Club of lending nations to eliminate all of its bilateral external debt. Under the agreement, the lenders will forgive most of the debt, and Nigeria will pay off the remainder with a portion of its energy revenues. Outside of the energy sector, Nigeria's economy is highly inefficient. Moreover, human capital is underdeveloped–Nigeria ranked 151 out of 177 countries in the United Nations Development Index in 2004–and non-energy-related infrastructure is inadequate. From 2003 to 2007, Nigeria attempted to implement an economic reform program called the National Economic Empowerment Development Strategy (NEEDS). The purpose of the NEEDS was to raise the country's standard of living through a variety of reforms, including macroeconomic stability, deregulation, liberalization, privatization, transparency, and accountability. The NEEDS addressed basic deficiencies, such as the lack of freshwater for household use and irrigation, unreliable power supplies, decaying infrastructure, impediments to private enterprise, and corruption. The government hoped that the NEEDS would create 7 million new jobs, diversify the economy, boost non-energy exports, increase industrial capacity utilization, and improve agricultural productivity. A related initiative on the state level is the State Economic Empowerment Development Strategy (SEEDS).

A longer-term economic development program is the United Nations (UN)-sponsored National Millennium Goals for Nigeria. Under the programme, which covers the years from 2000 to 2015, Nigeria is committed to achieve a wide range of ambitious objectives involving poverty reduction, education, gender equality, health, the environment, and international development cooperation. In an update released in 2004, the UN found that Nigeria was making progress toward achieving several goals but was falling short on others. Specifically, Nigeria had advanced efforts to provide universal primary education, protect the environment, and develop a global development partnership. However, the country lagged behind on the goals of eliminating extreme poverty and hunger, reducing child and maternal mortality, and combating diseases such as human immunodeficiency virus/acquired immune deficiency syndrome (HIV/AIDS) and malaria.

A prerequisite for achieving many of these worthwhile objectives is curtailing endemic corruption, which stymies development and taints Nigeria's business environment. President Olusegun Obasanjo's campaign against corruption, which includes the arrest of officials accused of misdeeds and recovering stolen funds, has won praise from the World Bank. In September 2005, Nigeria, with the assistance of the World Bank, began to recover US$458 million of illicit funds that had been deposited in Swiss banks by the late military dictator Sani Abacha, who ruled Nigeria from 1993 to 1998. However, while broad-based progress has been slow, these efforts have begun to become evident in international surveys of corruption. In fact, Nigeria's ranking has consistently improved since 2001 ranking 147 out of 180 countries in Transparency International's 2007 Corruption Perceptions.

20.3. Macro-economic Trend

This is a chart of trend of gross domestic product of Nigeria at market prices estimated by the International Monetary Fund with figures in millions of Nigerian Naira.

Year	*Gross Domestic Product*	*US Dollar Exchange*	*Inflation Index (2000=100)*	*Per Capita Income (as % of USA)*
1980	50,849	0.78 Naira	1.30	7.22
1990	286,374	8.94 Naira	8.10	1.49
2000	4,676,394	102.24 Naira	100	1.11
2005	14,894,454	131.01 Naira	207	1.96
2010	32,675,202	164.08 Naira	502	3.14

Source: World Bank Report, 2013.

For purchasing power parity comparisons, the US Dollar is exchanged at 75.75 Nigerian Naira only. Current GDP per capita] of Nigeria expanded 132% in the Sixties reaching a peak growth of 283% in the Seventies. But this proved unsustainable and it consequently shrank by 66% in the Eighties. In the Nineties, diversification initiatives finally took effect and decadal growth was restored to 10%.

Due to inflation, per capita GDP today remains lower than in 1960 when Nigeria declared independence. About 57 per cent of the population lives on less than US$1 per day. In 2005 the GDP was composed of the following sectors: agriculture, 26.8 per cent; industry, 48.8 per cent; and services, 24.4 per cent. In 2005 Nigeria's inflation rate was an estimated 15.6 per cent. Nigeria's goal under the National Economic Empowerment Development Strategy (NEEDS) program is to reduce inflation to the single digits. In 2005 Nigeria's central government had expenditures of US$13.54 billion but revenues of only US$12.86 billion, resulting in a budget deficit of 5 per cent. Nigerian tax authorities face the challenge of widespread tax evasion, which is motivated by complaints about corruption and the poor quality of services. Mean wages were $1.11 per manhour in 2009.

Nigeria ranks twenty fifth worldwide and first in Africa in farm output. Agriculture has suffered from years of mismanagement, inconsistent and poorly conceived government policies, and the lack of basic infrastructure. Still, the sector accounts for over 26.8% of GDP and two-thirds of employment. Nigeria is no longer a major exporter of cocoa, groundnuts (peanuts), rubber, and palm oil. Cocoa production, mostly from obsolete varieties and overage trees, is stagnant at around 180,000 tons annually; 25 years ago it was 300,000 tons. An even more dramatic decline in groundnut and palm oil production also has taken place. Once the biggest poultry producer in Africa, corporate poultry output has been slashed from 40 million birds annually to about 18 million. Import constraints limit the availability of many agricultural and food processing inputs for poultry and other sectors. Fisheries are poorly managed. Most critical for the country's future, Nigeria's land tenure system does not encourage long-term investment in technology or modern production methods and does not inspire the availability of rural credit.

Agricultural products include cassava (tapioca), corn, cocoa, millet, palm oil, peanuts, rice, rubber, sorghum, and yams. In 2003 livestock production, in order of metric tonnage, featured eggs, milk, beef and veal, poultry, and pork, respectively. In the same year, the total fishing catch was 505.8 metric tons. Roundwood removals totaled slightly less than 70 million cubic meters, and sawnwood production was estimated at 2 million cubic meters. The agricultural sector suffers from extremely low productivity, reflecting reliance on antiquated methods. Although overall agricultural production rose by 28 per cent during the 1990s, per capita output rose by only 8.5 per cent during the same decade. Agriculture has failed to keep pace with Nigeria's rapid population growth, so that the country, which once exported food, now relies on imports to sustain itself Nigeria ranks 44th worldwide and third in Africa in factory output. The oil boom of the 1970s led Nigeria to neglect its strong agricultural and light manufacturing bases in favor of an unhealthy dependence on crude oil. In 2000, oil and gas exports accounted for more than 98% of export earnings and about 83% of federal government revenue. New oil wealth, the concurrent decline of other economic sectors, and a lurch toward a statist economic model fueled massive migration to the cities and led to increasingly widespread poverty, especially in rural areas. A collapse of basic infrastructure and social services since the early 1980s accompanied this trend. By 2000, Nigeria's per capita income had plunged to about one-quarter of its mid-1970s high, below the level at independence. Along with the endemic malaise of Nigeria's non-oil sectors, the economy continues to witness massive growth of "informal sector" economic activities, estimated by some to be as high as 75% of the total economy.

Nigeria's proven oil reserves are estimated to be 35 billion barrels (5.6×10^9 m^3); natural gas reserves are well over 100 trillion cubic feet (2,800 km^3). Nigeria is a member of the Organization of Petroleum Exporting Countries (OPEC), and in mid-2001, its crude oil production was averaging around The types of crude oil exported by Nigeria are Bonny light oil, Forcados crude oil, Qua Ibo crude oil and Brass River crude oil. Poor corporate relations with indigenous communities, vandalism of oil infrastructure, severe ecological damage, and personal security problems throughout the Niger Delta oil-producing region continue to plague Nigeria's oil sector. Efforts are underway to reverse these troubles. In the absence of government programs, the major multinational oil companies have launched their own community development programs. A new entity, the Niger Delta Development Commission (NDDC), has been created to help catalyze economic and social development in the region. Although it has yet to launch its programs, hopes are high that the NDDC can reverse the impoverishment of local communities. The U.S. remains Nigeria's largest customer for crude oil, accounting for 40% of the country's total oil exports; Nigeria provides about 10% of overall U.S. oil imports and ranks as the fifth-largest source for U.S. imported oil.

The United Kingdom is Nigeria's largest trading partner followed by the United States. Although the trade balance overwhelmingly favors Nigeria, thanks to oil exports, a large portion of U.S. exports to Nigeria is believed to enter the country outside of the Nigerian government's official statistics, due to importers seeking to avoid Nigeria's excessive tariffs. To counter smuggling and under-invoicing by importers, in May 2001, the Nigerian government instituted a full inspection program for all imports, and enforcement has been sustained. On the whole, Nigerian high tariffs and non-tariff barriers are gradually being reduced, but much progress remains to be made. The government also has been encouraging the expansion of foreign investment, although the country's investment climate remains daunting to all but the most determined. The stock of U.S. investment is nearly $7 billion, mostly in the energy sector. Exxon Mobil and Chevron are the two largest U.S. corporations in offshore oil and gas production. Significant exports of liquefied natural gas started in late 1999 and are slated to expand as Nigeria seeks to eliminate gas flaring by 2008.

Oil dependency, and the allure it generated of great wealth through government contracts, spawned other economic distortions. The country's high propensity to import means roughly 80% of government expenditures is recycled into foreign exchange. Cheap consumer imports, resulting from a chronically overvalued Naira, coupled with excessively high domestic production costs due in part to erratic electricity and fuel supply, pushed down utilization of industrial capacity to less than 30%. Many more Nigerian factories would have closed except for relatively low labor costs (10%–15%). Domestic manufacturers, especially pharmaceuticals and textiles, have lost their ability to compete in traditional regional markets. However, there are signs that some manufacturers have begun to improve competitiveness. In April 2006, Nigeria became the first African country to fully pay off its debt owed to the Paris Club. However, this was structured as a debt writeoff of approximately $18 billion and a cash payment of approximately $12 billion.

In the light of highly expansionary public sector fiscal policies in 2001, the government sought ways to head off higher inflation, leading to the implementation of stronger monetary policies by the Central Bank of Nigeria (CBN) and underspending of budgeted amounts. As a result of the CBN's efforts, the official exchange rate for the Naira has stabilized at about 112 Naira to the dollar. The combination of CBN's efforts to prop up the value of the Naira and excess liquidity resulting from government spending led the currency to be discounted by around 20% on the parallel (non-official) market. A key condition of the Stand-by Arrangement has been closure of the gap between the official and parallel market exchange rates. The Inter Bank Foreign Exchange Market (IFEM) is closely tied to the official rate. Under IFEM, banks, oil companies, and the CBN can buy or sell their foreign exchange at government influenced rates. Much of the informal economy, however, can only access foreign exchange through the parallel market. Companies can hold domiciliary accounts in private banks, and account holders have unfettered use of the funds.

Expanded government spending also has led to upward pressure on consumer prices. Inflation which had almost disappeared in April 2000 reached 14.5% by the end of the year and 18.7% in August 2001. In 2000, high oil prices resulted in government revenue of over $16 billion, about double the 1999 level. State and local governments demanded access to this "windfall" revenue, creating a tug-of-war between the federal government, which sought to control spending, and state governments desiring augmented budgets, preventing the government from making provision for periods of lower oil prices.

Nigeria ranks 63rd worldwide and fifth in Africa in services' output. Low power and telecom density has crippled the growth of this sector. Since undergoing severe distress in the mid-1990s, Nigeria's banking sector has witnessed significant growth over the last few years as new banks enter the financial market. Harsh monetary policies implemented by the Central Bank of Nigeria to absorb excess Naira liquidity in the economy has made life more difficult for banks, some of whom engage in currency arbitrage (round-tripping) activities that generally fall outside legal banking mechanisms. Private sector-led economic growth remains stymied by the high cost of doing business in Nigeria, including the need to duplicate essential infrastructure, the threat of crime and associated need for security counter measures, the lack of effective due process, and nontransparent economic decisionmaking, especially in government contracting. As of 2007, 29% of Nigerians in urban areas did not own bank accounts. While corrupt practices are endemic, they are generally less flagrant than during military rule, and there are signs of improvement. Meanwhile, since 1999 the Nigerian Stock Exchange has enjoyed strong performance, although equity as a means to foster corporate growth remains underutilized by Nigeria's private sector.

Nigeria's publicly owned transportation infrastructure is a major constraint to economic development. Principal ports are at Lagos (Apapa and Tin Can Island), Port Harcourt, and Calabar. Docking fees for freighters are among the highest in the world. Of the 80,500 kilometers (50,000 mi.) of roads, more than 15,000 kilometers (10,000 mi.) are officially paved, but many remain in poor shape. Extensive road repairs and new construction activities are gradually being implemented as state governments, in particular, spend their portions of enhanced government revenue allocations. The government implementation of 100% destination inspection of all goods entering Nigeria has resulted in long delays in clearing goods for importers and created new sources of corruption, since the ports lack adequate facilities to carry out the inspection. Four of Nigeria's airports–Lagos, Kano, Port Harcourt and Abuja–currently receive international flights. Government-owned Nigeria Airways ceased operations in December 2002. Virgin Nigeria Airways started operations in 2005 as a replacement and serves domestic and international routes. There are several domestic private Nigerian carriers, and air service among Nigeria's cities is generally dependable. The maintenance culture of Nigeria's domestic airlines is not up to internationally accepted standards.

Although Nigeria must grapple with its decaying infrastructure and a poor regulatory environment, the country possesses many positive attributes for carefully targeted investment and will expand as both a regional and international market player. Profitable niche markets outside the energy sector, like specialized telecommunication providers, have developed under the government's reform program. There is a growing Nigerian consensus that foreign investment is essential to realizing Nigeria's vast but squandered potential. European investments are increasing, especially since Belgian consultancy companies such as Genco are exploring the Nigerian market. Companies interested in long-term investment and joint ventures, especially those that use locally available raw materials, will find opportunities in the large national market. However, to improve prospects for success, potential investors must educate themselves extensively on local conditions and business practices, establish a local presence, and choose their partners carefully. The Nigerian Government is keenly aware that sustaining democratic principles, enhancing security for life and property, and rebuilding and maintaining infrastructure are necessary for the country to attract foreign investment. The stock market capitalisation of listed companies in Nigeria was valued at $97.75 billion on February 15, 2008 by the Nigerian Stock Exchange.

In 2007, Nigeria received a net inflow of US$5.2 billion of foreign direct investment (FDI), much of which came from Nigerians in the diaspora. Most FDI is directed toward the energy and banking sectors. As of October 2005, World Bank assistance to Nigeria involved 19 active projects with a total commitment value of about US$1.87 billion. Since Nigeria joined the World Bank in 1961, the World Bank has assisted it on 120 projects. In October 2005, the International Monetary Fund approved a two-year "policy support instrument" designed to promote the growth of the non-oil sector and to reduce poverty. The United States assisted with Nigeria's economic development from 1954 through June 1974, when concessional assistance was phased out because of a substantial increase in Nigeria's per capita income resulting from rising oil revenue. By 1974, the United States had provided Nigeria with approximately $360 million in assistance, which included grants for technical assistance, development assistance, relief and rehabilitation, and food aid. Disbursements continued into the late 1970s, bringing total bilateral economic assistance to roughly $445 million.

The sharp decline in oil prices, economic mismanagement, and continued military rule characterized Nigeria in the 1980s. In 1983, USAID began providing assistance to the Nigerian Federal and State Ministries of Health to develop and implement programs in family planning and child survival. In 1992, an HIV/AIDS prevention and control program was added to existing health activities. USAID committed $135 million to bilateral assistance programs for the period of 1986 to 1996 as Nigeria undertook an initially successful Structural Adjustment Programme, but later abandoned it. Plans to commit $150 million in assistance from 1993 to 2000 were interrupted by strains in U.S.-Nigerian relations over human rights abuses, the failed transition to democracy, and a lack of cooperation from the Nigerian Government on anti-narcotics trafficking issues. By the mid-1990s, these problems resulted in the curtailment of USAID activities that might benefit the military Government. Existing health programs were re-designed to focus on working through grassroots Nigerian non-governmental organizations and community groups. As a response to the Nigerian military government's plans for delayed transition to civilian rule, the Peace Corps closed its programme in Nigeria in 1994.

In response to the increasingly repressive political situation, USAID established a Democracy and Governance (DG) program in 1996. This program integrates themes focusing on basic participatory democracy, human rights and civil rights, women's empowerment, accountability, and transparency with other health activities to reach Nigerians at the grassroots level in 14 of Nigeria's 36 states. The sudden death of General Sani Abacha and the assumption of power by General Abdulsalami Abubakar in June 1998, marked a turning point in U.S.-Nigerian relations. USAID provided significant support to the electoral process by providing some $4 million in funding for international election observation, the training of Nigerian election observers and political party polling agents, as well as voter education activities. A Vital National Interest Certification was submitted to Congress in February 1999 by President Clinton to lift restrictions on U.S. Government interaction with and support to the Government of Nigeria.

Since that time, USAID has supported Nigeria to sustain democracy and to improve governance by providing training on the roles and responsibilities of elected officials in a representative democracy for newly elected officials at the federal, state, and local levels prior to their installation in May 1999 and assisting with conflict prevention and resolution in the Niger Delta, civil military relations, civil society, and political party development. In the economic area USAID supports programs in strengthening economic management and coordination, encouraging private sector development and economic reform, helping Nigeria reap the benefits of AGOA, improved agricultural technology and marketing and smallscale and microenterprise development. In addition, health assistance, focusing on HIV/AIDS, nutrition, and immunization, education, transportation and energy infrastructure, are priorities for bilateral assistance.

References

"Doing Business in Nigeria 2012". World Bank. http://www.doingbusiness.org/data/exploreeconomies/nigeria/. Retrieved 2011-11-21.

"Sovereigns rating list". Standard & Poor's. http://www.standardandpoors.com/ratings/sovereigns/ratings-list/en/eu/?subSectorCode=39. Retrieved 26 May 2011.

"Motivation and barriers to learning for young people not in education, employment or training". A 2013 publication by the Department for Business, Innovation & Skills, London.https://www.gov.uk/government/publications/motivation-and-barriers-to-learning-for-young-people-not-in-education-employment-or-training

Abel, A. (1984) "Empirical Investment Equations: An Integrated Approach". Presented at Carnegie.

A Cemoghi, Deron, Jounson, Simon, Robinson, James, Thaica Uroen, Y. (2003). Institution caused, macro-economic symptom volatility 'crises and growth journal of monetary Economics 50, 49-123 (January)

Abubakar, D. (2001). Globalization, social sciences and Nigeria in the 21 st century. *The Nigerian Social Scientist*, 4, 1.

Adebiyi, M.A., 2000. Financial Liberalization and Baking Fragility in Nigeria. An empirical investigation. First bank BiB Ann. Rev., 8(17): 59-72.

Adedipe, A., 2005. Building and sustaining corporate performance and growth in the capital market. Nigeria stock exchange annual.

Adegbite, E.O, and S.I. Owualah (2007) "1st National Finance and Banking Conference on Economic Reforms and The Nigerian Financial System", A

Communique read at National Conference organised by the Department of Finance, FBA University of Lagos, Nigeria, 3-4, April, 2007.

Adegbite, I. (1992). "Effective growth and survival of small and medium-scale enterprises in Nigeria in the 1990s and beyond: the role of policy". In Ade, T.O. (1995): "Management of small and medium scale enterprises in Nigeria."Puncmac Publishers, Lagos.

Ademuwagun, J. (2011). "Who is Lee Kwan Yew?". Thought Revolution (Tr) Magazine, Volume 1, No. 1.

Aderibigbe, J.O. 1997. "Monetary policy and Financial sector Reform" CBN Bullion, vol. 21, No. 4, pp. 7-13.

Agarwal, J.P. (1980) "Determinants of Foreign Direct Investment: A Survey", *Weltwirtschaft-liches Archiv*, Vol. 117, pp.30-64.

Agenda: An Address by CBN Governor to the Nigerian Banker Committee.

Aghion, P., Benerjee A., Piketty, T. (1999), Dualism and neuroeconomic volatility. *Quarterly Journal of Economic* 14, 1397 (November).

Agu, C.C. (1988) *Nigerian Banking Structure and Performance: The Banking System's Contribution to Economic Development*, Onitsha: African-Fep Publishers

Aigbokhan, B.E. 1995. Macroeconomics: Theory, Policy and Evidence City: Idelojie publishers.

Aina, T. A. (1996). Globalization and social policy in Africa. *CODESRIA Bulletin*, 4.

Ajakaiye, O. 2002. "Economic Development inNigeria: A Review of Experience" CBN Bullion, Vol. 26, No. 1, pp. 47-64.

Ake, C. (1995). The new world order: The view from Africa, in Hans-Henrik and G Sovensen (eds), *Whose world order: Uneven globalization and end of coldwar*, London, West View Press.

Akinbobola, A. (1999). Globalization and economic development of emergent nation states: A Comparative study of Thailand, Kenya and Taiwan. *1998/99 MILD Journal and Year book*. Lagos.

Akinlami, T. (2011). "African Citizenship Issues". Thought Revolution (Tr) Magazine, Volume 1, No. 1.

Akinrinde, I. (2011) (Unpublished)."Transforming BRICS to BRINCS: thoughts on positioning Nigeria as a major world economic player."Nigerian Economics Students' Association (NESA), Unilag Chapter.

Akpan H. Ekpo The Economy and Jonathan's Transformation Agenda TELL, Wednesday, 28 September 2011.

Akpan H. Ekpo, Growth without Development in Nigeria, TELL, Thursday, 29, March 2010

Akpokodje, G. (1998) "Macroeconomic Policies and Private Investment in Nigeria", *In: Rekindling Investment for Economic Development in Nigeria* by Ben. E.

Aigbokhan (ed) Selected Papers for the 1998 Annual Conference, The Nigerian Economic Society.

Amsden, Alice (2001) : The rise of the rest: challenges to the west from late-industrializing Economics, oxford university press

an Integrated Supply Chain: Available from: http://www.mcb.co.uk.html

Anao. A.R., 1995. The Accountancy Profession and Nigeria economic development Bank overview. The Nigeria Accountant Official J. ICAN. ISSN 0048-0371. CBN, 2005. Central Bank of Nigeria annual report and accounts.

Andre S.J. Escrband, A., Molina, C. and Taguas, D. La Inversun en Espana Econometria con REstricaiones de EQuilibro. Barcelona: A Bosh, 1990.

Anya O. A. (2001,May 7). The challenge before the economy. *The Post Express*, Lagos.

Anyanwu, J.C. 1993. Monetary Economics: Theory, Policy and Institutions, Onitsha: Joanee Educational Publishing Ltd.

Appadurai, A. (2001). Grassroots globalization and the research imagination, in A.Appadurai, (ed), *Globalization*, Durham: Duke University Press.

Aremoghu, D. (1993). "Learning About Other's Actions and the investment Accelerator." *The Economic Journal* 1993, 103 (417), pp. 318-328.

Aremu, J.A. (1997) "Foreign Private Investment: Issues, Determinants, Performance and Promotion", *CBN Research Seminar Papers*, No. 4 (July, 1997)

Arestis P., Dennlnades, and Luintel K, (2001). Financial Development and Economic Growth: The Role of stock markets, Journal of money, credit and Banking, 33(1) 16-41 in Olofin, S.O. and Afangidon, U.J. financial structure and Economic Grwoth in Nigeria. A Macro-Economic Approach.

Arestis, P.,Dennelmades and Luintel, k.(2001) financial development and Economic growth: the rule of stock market. *Journal of Money and Banking.*

Aryeetey, Ernest (2003): Resent Development in African financial market: Agenda for further Research: *Journal of African Economics* 12 sup. 2 :111-ii152

Asian development Bank (2000): The role of central banks in microfinance in Asia and the pacific. ADB.

Available from:http://www.mcb.co.uk.html (Accessed on 13 March 2002).

Azam, Jean-Paul, Augustin Fosu, Njuguna S. Ndung'u (1999) "Explaining slow growth in Africa". African Development Review, Volume 14, Issue 2,

Bacha, E. (1984) "Growth with Limited Supplies of Foreign Exchange: A Reappraisal of the Two Gap Model", The World Bank.

Baddeley, M. (2003). Investment: Theories and Analysis. Basingstoke: Palgrave Macmillan.

Ball, 2 and Sheridan, N (2003) "Does Inflation Targeting Matter?" IMF working paper No.03-129.

Beck, U. and Beck, G. (2002). Individualization. London Sage.

Bekaert, G., Harvey, C., Lundblad, C. (2002). Does financial liberalization Spur Growth? Unpublished manuscript, Columbia, University, New York.

Bekaert, G., Harvey, C., Lundblad, C. (2002). Growth volatility and equality market liberalization Duke University working paper.

Bermamke, Bins, S. Thomas Laubach, Adams Posen and Federic S. misukin (1999). Inflation Targeting : lesson from the intermatiional Experience. Plrlillnceton NJ: Princeton University Press.

Blejer, m, Leome, A.M, Rebanal, P. and Schwartz G. (2001). Inflation targeting in the context of IMF adjustment programs, international monetary fund working paper no. 01/31.

Blejer, Mario I., Alfredo M. Leone, Pau Rabanal and Gerd Schwartz, 2002. "Inflation targeting in the Contect of IMF-Supported Adjustment Programs", IMF staff papers. Vol. 49, no. 30 pp. 313-338.

Book Factor, R. (2001). Supply Chain Management. *The Retail Advisor Fall.* 2001:7.

Bouce James and Leonce Ndikumana (2001) "is Africa a net creditor? New Estimate of capital Flight for severally indebted Sub Saharan African countries 1970-1996, *Journal of Development Studies*, 38(2), 27-56.

Braddy, B. (2002). New Avenues for Logistics Cost Reductions: Replacing Inventory With Information. *Inbound Logistics* Jan. 2002: 14-15.

Braunstein, Elissa and James Heintz (2006) "Gender Bias and central Bank policy: employment and inflation Reduction". *International review of Applied Economics* March, 2008.

Bruno, M. and W. Easterly (1996). Inflation and Growth: In search of a stable Relationship. *Federal Reserve Bank of St. Louis Review*, 78(3), pp. 139-46.

Bruno, Michael and William Easterly. 1998. "Inflation and Growth: in Search of a stable Relationship, "*Federal Reserve Bank of St. Louis Review*, 78 (3), pp. 139-146.

Bruno, Micheal and William R. Easterly (1998), "inflation crises and long-run growth". ' *Journal of monetary Economics* 41, 3-26.

Buch, C.M., Dopke, J., Plerdzioch C.I. (2002). Financial openness and Business cycle volatility, *Kid Institute for World Economics Working Paper.*

Byrn, R.T. and G.W. Stone 1992. Economics, U.S.A: Harper Collion.

Cadbury, A., 2002. A Corporate Governance and Chairmanship. Oxford University Press. *J. Bean, Thear,* 2 (3): 112-117, *2008.*

Calstrom, C. and G. Samolysk. 1995. Loan sales response to market based capital requirement. *J. Banking Finance*, 19: 627-646.

Camara S. (2008). "Determinants Reales financies de la Inversion en Esiana cion unidos Analsis. Teorico Y estimacion Empirica, Bilbao: Xi Jornadas de *Economica Critica* 2008.

Carver, T.N. (1903). "Therelation of Abstinence to interest" *Quarterly Journal of Economics*, 18(1) pp. 142-145.

CBN Economic and financial review, 1981, 1986.

Center for Global Development : Publications: Resolving Nigeria's Debt Through a Discounted Buyback.

Chenery, H. and Bruno (1962) "Development Alternatives in an Open Economy: The Case of Israel", *Economic Journal*, 72.

Chenery, Hollis, B. and Monsue Syrquin (1975). Patterns of Development, 1950-1970 Oxford Economics paper, 50, 563, 72.

Chete, L.M. NI (1998) "Determinants of Foreign Direct Investment in Nigeria", *SER MONOGRAPH SERIES* No. 7, (1998), pp.289-336.

Chhiber, A and M. Dailami (1990) "Fiscal Policy and Private Investment in Developing Countries: Recent Evidence on Key Selected Issues" *PPR Working Paper*, No. 559, The World Bank.

Chirinko, Robert (1993). "Business fixed investment spending: Modeling Strategies, Empirical Results and Policy Implications". *Journal Economic Literature*, 31 (December, 1875-1911).

Chopra, S. and Meindl, P. (2001). *Supply Chain Management: Strategy, Planning and.*

Clarinda R., Gali J. and Gerther, M. (1999) "The Science of Monetary policy: A new Keynesian perspective". Journal of Economic literature 1999, 37(4), 1661-1707.

Clark, I. (1999). *Globalization and international relations theory*, Oxford: Oxford University Press.

Clark, J.M. (1917). "Business Acceleration and law of demand. *Journal of Political Economy.*

Cooke, J.A. (2000). Logistics ropes in Inventory. *Management and Distribution Report.* July.

Coyle, J.J. Bardi, E.J. and Langley, C.J. (1996). The Management of Business Logistics: 6th.

Crainic, G.T. (2002).A survey of optimization models for Long-Haul freight Transportation.

Davis, P. and Drum,F. (1996). *Council of Logistics Management Annual Conference*, Oak.

Dejuan, O. (2005) "Paths of Accumulation and Growth: Towards a Keynesian Long-period theory of output. "Review of Political Economy, 2005. 17 (2), 231-252.

DFID Policy Paper Dec. (2005) Growth and Poverty Reduction: The Role of Agriculture Published by the Department for International Development.

Dunning, J.H. (1973). "The Determinants of International Production", *Oxford Economic Papers* Vol. 25, (November), pp.289-336.

Dunning, J.H. (1981). "Explaining the International Direct Investment Position of Countries: Towards a Dynamic or Developmental Approach" *Wiltwirt-schaftliches Archiv* Vol.117, pp.30-64.

Easterly W., Islam, R.J., Stiglitz, J.E. (2001). Suaken and stirred:" Explaining growth volatility in: Pleskovik B., Stern, N. (Eds). Annual World Bank Conference on Developing Economics.

Easterly, William (2002). "An Indentity Crisis? Testing IMF Financial Programming". Center for Global Development. Working Paper No. 2, August.

Edozien, E.C. (1968). "Linkages: Direct Foreign Investment and Nigeria's Economic Development", *Nigerian Journal of Economic and Social Studies* Vol. 10(2), pp.191-203.

Ekpo, A.H. (1997) "Foreign Direct Investment in Nigeria: Evidence from Time Series Data", *CBN Economic and Financial Review*, Vol. 35, No. 1.

Energy Information Agency - Data and Analysis of Oil Production.

Epstein, Gerald (2007a). "Central bank as agents of economic development", in Ha-Joon chang,ed. Institutional change the economic development, 2007, united nations, university and Anthem press.

Epstein, Gerald, 2003. "Alternatives to Inflation Targeting Monetary Policy for stable and Egalitarian Growth: A Brief Research Summary. PERI Working paper. No. 62, www.umass.edu/peri>.

Espasa, A. and SEnra, E. (1993). "Consideraciones sobre ia function de inversun en Espana working paper, 1993.

Falemi-Ozcan, S., Sorensen B.E., Yosna, O. (2003). Risk-sharing and industrial specialization regional and International audience. American Economic Review as, 903-918 (June).

Fatas, A. (2002). The effect of Business cycles on growth in Loayza, N. Soto, R.C. (ed) Economic Grwoth Sources, Treadsanol cycles, central Bank of chille, Santiago.

Fawcett, P. Mcleish, R. and Ogden, I. (1992). *Logistics Management*, Singapore: Financial.

Fazzari, (1987). S.M. and Mott, T.L. "The Investment Theories of kalacki and Keynes: An Empirical study of firm Data, 1970-1982", *Journal of Post Keynesian Economics Winter* 1986-87, 9(2), pp. 171-187.

Fazzari, G., R.G. Hubbard and B.C. Petersen (1988a). "Financing Constraints and Corporate Investment", *Brooking Papers on Economic Activity*, Vol. 1 No. 2.

Fazzari, G., R.G. Hubbard and B.C. Petersen (1988b). "Investment, Financing Decisions and Tax Policy". *American Economic Review*, 82(3).

Federal Republic of Nigeria: National Rolling Plan, 1991-93.

Filho, Alfredo Saad (2004). "Pro-poor Monetary and Anti-Inflation Policies Developing Alternatives to the New Monetary Policy Consensus". Mimeo. SOAD, University of London.

Fisher I. (1930). The theory of interest, New York. Macmillian 1930.

Fisher, S. (1993). "The Role of Macroeconomic Factors in Growth", Journal of *Monetary Economics* Vol. 32 (December) pp. 458-512.

Frenceschin, F. and Rafele, C. (2000). Quality evaluation in logistics services. *International.*

Frenkel, Reberto and Rapetti, martin {2008). "Five years of competitive and stable real exchange rate in Argentina, 2002-2004" international Review of Applied Economics.

Frenkel, Roberto and Martin Rapetti (2008). "Monetary and Exchange rAte Policies in Argentina after the Convertibility Regime Collapse", Intentional Review of Applied Economic March.

from:http://www.ascet.com/html (Accessed on 26 September 2003).

Fukuyama, F. (1992). *The end of history and the last man*, London: Hamish Hamilton.

Giblin, J. (2001). *Transportation: The Forgotten link in the Supply Chain*: Available.

Gilbert Gallopin (1994). Impoverishment and Sustainable Development: A System Approach. A report of the International Institute for Sustainable Development, IISD, Manitoba, Canada.

Gilmour, P. (1997). Information flow in materials management. *International Journal.*

Granger, C.W.J. (1986). "Developments in the Study of Co-integrated Variables", *Oxford Bulletin of Economics and Statistics*, Vol. 48, pp.213-228.

Granger. C.W.J. 1969. An Investigating Causal Relations by Econometric Models and Cross Spectral methods. ECONOMETRICAL, 37:428-438.

Green, Anke (2003). "Credit Guarantee schemes for small scale Enterprises: An effective instrument to promote private sector- led Growth" UNDO working paper, SME technical working paper series.

Greene, J. and D. Villannueva (1991). "Private Investment in Developing Countries: An Empirical Analysis", *IMF Staff Papers* Vol. 38, No. 2 (March) pp.33-58.

Greenspan. A. 2004. A Remarkl at the Conference on Bank Structure and Com pennon. Sponsored by the Federal Reserve Bank of Chicago Illinois via Satellite.

Greenwood, J. and R. Jovanovic (1990). "Financial Development, Growth and Distribution of Income", *Journal of Political Economy*, Vol. 98, pp.1076-107.

Greunning, H.V. and S.B. Bratanovic, 2003. A Analysis and Managing banking Risk:, A Framework for Assessing Corporate Governance and Financial Risk. 2nd Edn. International Bank for Reconstruction and Development/the World Bank 1818 H Street N.W Washing DC 20433.

Hadjmichael, M., D. Ghura, M. Muhleisen, R. Nord and E. Ucer (1995). *Growth, Savings and Investment*, 1986-93. New York: International Monetary Fund.

Handbook of Transportation Science. R.W. Hall 2nd Edition, Vol 15,102-122. U.S.A: Kluwer.

Harrison, Ann. 1996. "Openness and Growth: a Time - Series, Cross-Country Analysis for Developing Countries". *Journal of Development Economics*, 48. Pp. 419-447.

Hayashi, F. (1982) "Tobin's Marginal Q, and Average Q: A Neoclassical Integration", *Econometrica* Vol. 50, No. 1.

Hnatkouska, V., Loayza, N. (2005). Volatility and growth in Alzenman, J., Pinto, B. (eds), Managing Economic volatility and crises: A practitioner's guide. Cambridge University Press.

Homohan and O' Connell(1997) "contrasting monetary regimes in Africa", IMF working paper WP/97/64 may.

http://www.blackwell-synergy.com/doi/pdf/10.1111/1467-8268.00051 (accessed 05.11.2007.

http://www.infochain.org.htlm (Accessed on 26 September 2003).

Ibrahim, J. (2002). Notes on globalization and the marginalisation of Africa. *CODESRIA Bulletin*, 3 and 4.

Ikuomola, Y. (2012)."Volunteering: standing up to be counted in the task of nation-building...".Venture Career Journal. pp. 92-93.

IMF (2002). World Economic Outlook International Monetary fund, Washington DC. In: Nnadi, K.U. and Ogwude I.C (ed) Transportation in Nigeria: Economics, Policy

International Monetary Fund (2006). "Inflation Targeting and the IMF" by Nicolatia Batini, Peter Bever Kalpana. Kochhar and Scott Rogar). www.imf.org.

Isijola, CO., 2000. Impact of financial sector, Reform on the supply and Demand For Agricultural Credit in Nigeria. *First Bank Bio Ann. Rev.*, 8:16.

Iyoda M. (2005). "The Determination of Investment in monetary production economy: a theory and it Empirical Application in the post Keynesian Tradition". *Journal of post Keynesian Economics*, fall 2005.

Jayedau, Arjun, (2006). "Dittering preferences between anti-inflation and anti-unemployment policy among the rich and poor: *Economics Letters* 91. pp. 67-71.

Jayeday, Argun (2008). "The class content of preferences toward Anti-inflation and Anti-unemployment policies". *International Review of Applied Economics*. March.

Jones, R.C. (1998). "Introduction: The Renewed Role of Remittance in the New Order", *Economic Geography* 74C17, 1-7.

Jorgenson, D.W. (1963). Capital theory and investment behaviour" *American Economic Review*, 1963, 53(2), pp. 247-59.

Journal of Agile management systems. 2(1), 40-53.

Jovanovic B. (2004). Asymmetric cycles NBER working paper.

Karnaiah, A., 2001. A Monetary Policy, expected inflation real activity and stock returns in India: An empirical analysis. *Asia-Mr. J.Econometric*, pp: 191-200.

Katarzyna Karpowicz, (2008). "Economic Growth and Development: Is High GDP Enough? Aberystwyth University.

Kelecki, M. (1971). Selected Essays on Dynamics of Capitalist Economy 1933-1970, Cambridge: Cambridge University Press.

Keynes, J.M. (1936). *The General Theory of Employment, Interest and Money Volume III of the Collected Writings of John Maynard Keynes*, London: McMillan.

Khan, M. and C. Reinhart (1990). "Private Investment and Economic Growth in Developing Countries", World Development Vol. 18 (January), pp.19-27.

King, R.G. and R.E. Levine (1993). "Finance and Growth: Schumpeter might be right", *Quarterly Journal of Economics* Vol. 108 (August) pp.717-737.

Korsah. Kweku Baa, Ernest K. Nyarko and Noel A. Tagos, 2001. "Impact of Financial Sector Liberalisation on Competition and Efficiency in the Ghanaian Banking Industry". IFLIP Resreach paper 01-2 Geneva: ILO.

Kose, M. A. Meredith, G.M. Towe, C.M., in press a. How has NAFTA affected the Mexican economy? Review and evidence.

Kose, M.A., Pasad, E.S., Terrones, M.E. (2003). Financial integration and macro-economic isolatility. IMF staff paper 50, 119-142.

Kose, M.A., Pasad, E.S., Terrones, M.E. (2005). How trade and financial integration affect relationship between growth and volatility? IMF working paper 05/19.

Kurz, H. (1986). "Normal position and capacity utilization" political economy: Studies in the surplus Approach, 1986, 2, pp. 37-54.

Kwanashie, M. (1999). Concepts and dimensions of globalization, in Nigerian Economic Society, *Globalization and Nigeria's economic development*, Ibadan: NES.

Lambert, D.M. (2001). *Supply Chain Management*: What does it involve?

Lambert, D.M. and Stock, J.R. Ellram, L.M. (1998). *Fundamental of Logistics Management*.

Largley, K.M. (1968). "The External Resource Factor in Nigeria's Economic Development", *Nigerian Journal* 182 *of Economic and Social Studies* Vol. 10(2), pp. 151-190.

Levuve, R. (2002). Bank- based or market-based financial system, which is better? *Journal of Financial International* (UCH): 398-428.

Lewis, M. (1979). "Does Political Instability in Developing Countries Affect Foreign Investment Flow?: An Empirical Examination", *Management International Review*, Vol. 19, pp.59-68.

Lipalile, M. (2001). Globalization, regional integration and development in Southern Africa, in Eddy Maloka (ed), *A United States of Africa*, Pretoria: Africa Institute of South Africa.

Lunn, J. (1980). "Determinants of US Direct Investment in the EEC: Further Evidence",*European Economic Review*, Vol. 13, pp. 93-101.

Lyoha M.A. 1998. Macroeconomics for Developing World Benin City: Miyo Educational Publisher.

Madunagu, E. (1999, July 26). Globalization and its victims, in *The Guardian*, Lagos.

Malinvaud, E. (1980). *Profitability and Unemployment*. London: Cambridge University Press.

Malnvaud, E. (1982). "Wages and Unemployment", *Economic Journal*, 92(1).

Management. Reliable Publisher, Owerri, Nigeria.

Manufacturing companies in Southwestern Nigeria, 2002-2006, Unpublished Ph.D Thesis,.

Mckinley Temy (2006). "the monopoly of Global capital flow: who needs structural Adjustment Now?, UNDP, international poverty center, working paper ₦12.

McKinnon, R.I. (1973). *Money and Capital in Economic Development*, Washington DC: The Brookings Institutions".

Mcknnon, R., 1973. Money and 6apital in Economic development; Washington D.C The brooking institution. New York; Oxford University Press.

Miligan, B. (2000). *Transportation and the Supply Chain: Transportation still falls through.*

Minsky, H. Joun Mayn and Keynes, New York: Columbia, 1975.

Mishkin, F., 1992. The Economic of money, Banking and Financial Markets. 3rd Edn. Harper Collin Publishers New York.

Mishkin, Frederic, S. and Schmidt hebbel (2007). Does inflation Targeting make a difference? *HBER Paper*, 12876, (2007).

Mishkin, Trederic S. and Klaus S. Schmidst-Hebbel (2001). One Decade of inflation: Targeting in the world: what do we know and what do we need to know? NBER working paper 8397.

Mittleman J.H. (1996). How does globalization work? in Mittleman J. H (ed), *Globalization: Critical reflection*, Boulder, Colorado.

Mittleman, J. (2000). *The Globalization syndrome: Transformation and resistance*, Princeton: Princeton University Press.

Mukhejee, C, 2002. The nature of the Causal relationship Between Stock Market acroeconomic Aggregates in India: An Empirical Analysis. Panda, Chakradhara Department of Economics, Jadavpur University.

Nabudere, D. (2000). Globalization, the African post-colonial state, post traditionalism and the new world order, in D. Nabudere (ed), *Globalization and the post-colonial African state*, Harare: AAPS Books.

Nabudere, Dani W. (1997) 'Beyond modernization and development, or why the poor reject development', Geografiska Annaler, Series B: Human Geography, http://www.blackwell-synergy.com/doi/pdf/10.1111/j.0435-3684.1997.00019.x (accessed 10 November 2007).

National Bureau of Statistic (NBC) Nigeria Poverty Profile Report, 2010. Press Briefing by the Statistician-General of the Federation, 13th Feb. 2012.

National Planning Commission (2004). National Economy and Empowerment Strategies (NEEDS) Abuja, Nigeria.

NDIC, 1996. Nigeria Deposit Insurance Corporation Annual Reports and Accounts.

Ndikumana Leonce (2000). "Financial Determinations of Domestic Investment in Sub-Saharan Africa. Evidence from panel Data". World Development, 28(2), pp. 381- 400.

Ngozi Okonjo-Iweala and Philip Osafo-Kwaako (March 2007) Nigeria's Economic Reforms: Progress and Challenges The Brookings Institution 1775 Massachusetts Ave., NW Washington, DC 20036.

Ninsin K. A (2000). Globalization and the future of Africa. *AAPS Occasional paperseries*, 4, (No. 1).

Nissanke maehiko and Aryeeted, Ernest (1998). Financial integeration and development Liberalization and Reform in sub-saharan Africa, London and New-York : Routledge.

Nnanna, O.J and Dago, Mela (1998). Structure Reform, monetary policy and financial deeping: The Nigerian experience, economic and financial Review, vol.36 no.2, June pp. 1-29.

Nzott, S.M. (2004). Money, banking and finance, theory and practice Owerri. Huason Jude publisher.

Obadan, M.I. (1994). "Direct Investment in Nigeria: An empirical Analysis", *African Studies Review* Vol. XXV, NO. 1, pp.67-81.

Oboh, G.A.T., 2005. <a Globalization and Competitiveness; concept, Fears and Prospect For Common Wealth Countries. Union Digest, 9:384.

Ocampo, Jose Antonio (2002). "Capital Account a Counter-Cyclical Prudential Regulations in Developing Countries". WIDER Discussion Paper, Helsinki WIDER UNU.

Ocampo, Jose Antonio 2003. A capital Account and counter-cyclical prudential Regulation in Developing countries'' in form capital surges to Drought. Seeking stability for emerging Economic, edited by Richardo French –Dauis and stephany Gritith – Jones 217-44, New –York, palgrava Macmillian.

Odozi, V.A. (1995). "An Overview of Foreign Investment, 1960-65" CBN *Research Department Occasional Paper*, No. 11 (June).

Ofei, Kwadwo Ansah (2001). "Terms and Access to credit: Perceptions of SME/ Enterpreneurs in Ghana", University of Nigeria.

Office of the Secretary to the Government of the Federation (OSGF) 2012. Nigeria in Transformation: Executive Intelligent Management Course 9.

Ogamba, E.N. (2003). "The Impact of Globalisation on Foreign Capital Flows in Nigeria", An Unpublished Ph.D. Thesis submitted to the Department of Banking and Finance, UNN, Enugu Campus, Enugu State.

Ogun, T. P. and P.O. lyoha, 2005. The Nigerian Stock Market and Future Economic Activity: Does the Deregulation of Financial Markets make any Difference? Union Digest, 1:2.

Ogunleye, G. A., The causes of Bank Failure and Persistent Distress in the Banking Industry. NDIC Quarterly, Vol. 13.

Okagbue. S. and T, Aliko, 2005. Banking Sector Reforms in Nigeria International legal News, 2:2.

Olabisi Onabanjo University, Ago-Iwoye.

Olakampo, O. (1962). "Foreign Aid and the Plan", *Nigerian Journal of Economic and Social Studies*, 492), pp.116-125.

Olaloku, F.A. (1979). *Structure of the Nigerian economy*, London.

Olaniyi, O. (1988). "An Econometric Analysis of Domestic Savings and Investment Financing in Nigeria", *Journal of Teacher Education* Vol. 4, No. 1, pp.133-142.

Olofin, S.O. and Afandiegin, U.J. (2008). Financial structure and economic growth in Nigeria Journal of Security and finance vol.13 No 1.pp. 47-48.

Olopoenia, R.A. (1983). "Foreign Investment and Growth Rates of Capital -Importing Countries: A Technical Note", *Nigerian Journal of Economic and Social Studies*, Vol. 25, No. 2 (July) pp.221-232.

Onyekpe, N.J.G. (2001). Globalization and liberalization of the world economy: Concepts, historical development and implications for the less developed countries and labour movement, in Sylvester Odion-Akhaine (ed), *Globalization, United Nationsand the fringe players*, Lagos: CENCOD.

Oriakhi, F. (2001). Globalization as imperialism : The Nigerian experience, in Sylvester Odion-Akhaine (ed), *Globalization, United Nations and the fringe players*, Lagos: CENCOD.

Osaghae, E. (1998). *Crippled giant: Nigeria since independence*. London: C. Hurst and Company (Publishing) Limited.

Oshikoya, T. (1994). "Macroeconomic Determinants of Domestic Private Investment in Africa: An Empirical Analysis" *Economic Development and Cultural Change*, Vol. 42, No. 3.

Oyaide, W.J. (1977). The Role of Direct Private Foreign Investment in Economic *Development: A Case Study of Nigeria*, 1962-1973. Washington DC: United Press of America.

Oyesiku, O.O. (1995). *Understanding Multiple Regression Analysis*. Lagos:HEBS Publisher.

Pittman, K. (1993)."About promoting youth development in schools". Seminar with 4-H Faculty, St. Paul, University of Minnesota.

Pollin, R. and Zhu, A. (2009). Inflation and economics Growth: Across country Non-Linear Analysis. Journal of post kiymesian Economics summer 40128 No.4.

Pollin, Robert and Andong Ahu. 2005. "Inflation and economic growth: a cross-country non linear analysis, "PERI Working paper 109, Amherst, MA: political Economy Research Institute.

Rabiu, M. (2012)."Promoting ethics and organizational values in the work place". *Venture Career Journal*. pp. 94-95.

Ramey, G. and Ramey V.A. (1995). Cross-country evidence of the link between volatility and growth. American Economic Review 85, 1138-1151 (December.).

Rationing Model with Monopolistic Competition", *European Economic Review*, 3(2) NATIONAL BUREAU OF ECONOMIC RESEARCH.

Raymond, J.L., Maroto, J.A. and Melle, M. (1999). Inversion Crecimiento empresial: Factors conducionantes. "Peoples de Economia Espanola, 1999, 78-79, pp. 102-121.

Reinhart Carmen and kennetti Rogott (2008a). "This time is different. A panoramic view of Eight centuries of financial crises "NBER working papei 13882, March.

Reinhart, Carmen and Kcenneth Rogott (2008b). "Banking Crises: An equal opportunity menace". NBER, Working paper on: 14587. December.

Robinson, J.V. (1962). Essay on the theory of Economic Growth, London, Macmillan 1962.

Rochester Conference on Public Policy on the State of Economics.

Rogers, Simon; Sedghi, Ami (15 April 2011). "How Fitch, Moody's and S&P rate each country's credit rating". *The Guardian*. http://www.guardian.co.uk/news/datablog/2010/apr/30/credit-ratings-country-fitch-moodys-standard. Retrieved 31 May 2011. .

Samuelson, P (1939). "Interaction between the multiplier analysis and principle of acceleration. "Review of Economic statistics, 1939, 21 (2), pp. 75-78.

Sani, T.A. and R.F. Yakpogoro, 1997. The Impact of Banking reform on Commercial banks Performance in Nigeria. *J. Indus. Business.* (JTBER) 1:2.

Sannes, R., 2001. Self-Service Banking. Value Creation Models and Information. Exchange. Special Services on Information Exchange on Electronic Markets, 4:3.

Scapelanda, A., and R. Balough (1988). "Determinants of US Direct Investment in the EEC", *European Economic Review* Vol. 21, pp.381-390.

Scarf, H. and Shoven, J.B. (1984). Applied General Equilibrium analysis. Cambridge University Press.

Scholte, J. (2000). *Globalization : A critical introduction*, London: Macmillan Press.

Seabrook, Jeremy (1993). Victims of Development. Resistance and Alternatives. London, New York: Verso.

Serrano, F. "Long period Effective Demand and the Scattian supermultiplier" contribution to political economy 1995, 14, pp. 67-71.

Serven, L. and A. Salimano (1991). "Private Investment and Macroeconomic Adjustment: A Survey", *World Bank Research Observer* Vol. 7 No. 1.

Shaw, E. (1973). *Financial Deepening in Economic Development*, New York: Oxford University Press.

Sheikh, A. (1991). "Wandering around the warranted path: Dynamic Non-linear solution to the harrodian Knife-edge" in E.J. Heil and W. Semmler (eds) Nicholas Kalder and mainstream Economics: London: Macmillian, 1991, pp. 320-334.

Singu, T. (2009). "Testing the neoclassical long-run and the Keynesian short-run Effect of Investment on output and Growth in India. "*Journal of Post Keynesian Economics Winter* 2008-2009, 31(2), pp. 271-297.

Skjott-Larsen, T. (2000). Third-Party Logistics: From an Interorganisational Point of View.

Sneessens, H. (1987). 'Investment and the Inflation - Unemployment Trade - Off in a Macroeconomic.

Soludo, C.C. 2004. 13-Point Nigerian Bank Reform.

Somuyiwa, A.O (2007). "The Relationship between the concepts of Logistics and Distribution".

Somuyiwa, A.O. (2010). Analysis of Logistics cost in the Supply Chain Management.

Sowa, Nii K. 2002. "Financial Sector Reform Policies and Poverty Reduction, "Working paper Center for policy Analysis (CEPA), Accra, Ghana.

Stallings, Barbara and Rogerio student (2006). "Finance for development''. Washington, DC : Brooking institution and united nation ECLAC.

Stank, T.P. and Goldsbug, T.J. (2000). *A Framework for Transportation Decision Making*.

Stightz, Joseph E. Manson UJ (1966). Financial markets public policy and East Asian miracles World Bank Research observer, U. (2) 249-276.

Storey, Andy (2003) "Measuring Development", in McCann, Gerard and Stephen McCloskey (eds.) (2003) From the Local to the Global. Key Issues in Development Studies. London, Starling: Pluto Press.

Symonides, J. (ed). (1998). *Human Rights: New dimensions and challenges*, Dartmouth, Dartmouth Publishing Company.

Tandon, Y. (2000). Globalization and Africas options, in Dani W. Nabudere, (ed), *Globalization and the post-colonial African state*, Harare: AAPS.

The 2011 UN Human Development Report.

Titilola, S.O and Akanji B.O (2005) Strategies for Planning and Budgeting for Economic and Social Sectors in Nigeria. A Paper Presented At a Training Workshop on Seeds Planning and Budgeting Technique for Enhancing the Performance of Planning and Budgeting Officers in Lagos State.

Titilola, S.O. Managing the Nigerian Economy: Challenges of the National Economy and Empowerment Strategies (NEEDS). A Paper Presented At a Training Programme for the Nigerian Petroleum Development Company (NPDC) On Developing Managerial Skill.

Todaro, Michael P., Stephen C. Smith (2006). Economic Development. Harlow: Pearson Education Limited.

Tornell, A, WEster mann, F., Martinez, L. (2004). The positive work between financial liberalization, growth and crises. NBER working paper. 10293.

Toyo, E.(2000). Background to globalization. Ibadan: ASUU Publication series. UNDP, National human development reports for Nigeria, 2000/2001.

Trezzini, A (1995). "Capacity utilization in long-run and Autonomous components of Aggregate Demand." Contributions to political Economy, 1995, 14, pp. 33-36.

Trezzini, A. "Capacity utilization in long-run: Some further considerations contributions to political Economy, 1998, 17, pp. 53-69.

UNDP (2007). MDG in Nigeria: Current Progress, UNDP, Abuja.

Usman, S. (1999). Implications of globalization for the Nigerian economy, in Nigerian Economic Society, Globalization and Nigeria's economic development. Ibadan: NES.

Vavrus, J. & Fletcher, A. (2006). "Guide to social change led by and with young people." The Freechild Project.http://www.freechild.org/socialchangeguide.htm.

World Bank (1995). A Continent in Transition: Sub-Saharan Africa in the Mid 1990s. Africa Region, World Bank, 1995. Washington, D.C.

World Bank Institute (2008). Development Outreach, Vol. 10, No. 3.

Yaqub, N. (2003). Review of the 2000/2001 National human development report, in The Nigerian social scientist, 6,(No.1).

Yusuf, A. A. (1994). Africa: A case for mass democratic participation, in Omoruyi, Omo et al., Democratization in Africa : African Perspective. Abuja: C.D.S.

Index

www.ingramcontent.com/pod-product-compliance
Lightning Source LLC
Chambersburg PA
CBHW071241230926
51849CB00001B/184
* 9 7 8 9 3 8 6 0 7 1 5 1 4 *